UX

Simple and Effective Methods for Designing
Great Products Using UX Programming Theories

ERIC SCHMIDT

Table of Contents

Introduction

If you have picked up this book, it means you are interested in learning about UX programming theories and how you can use them to design great products. Trust us. This book will not disappoint!

You will be surprised at how much these theories can help you assess the strengths and weaknesses of your product, identify user needs, and generate new ideas for product development. User experience (UX) programming theories can be used to design great products. By using these theories, you can create products that are easy to use and that meet the needs of your users.

In this book, you will learn about a variety of UX programming theories and methods and how to use them to design great products. You will learn about user research, user testing, and how to use data to make decisions about your product. This book will also teach you how to apply these theories in a real-world setting.

By the end of this book, You will be able to apply these theories to your own work. It is especially geared towards beginners and those who are new to the field of UX programming. The book covers a wide range of topics, from the basics of UX programming to more advanced concepts.

We have covered the complete lifecycle of product development, from user research to user testing and analysis. To make the book interesting to read, we have divided the product process phases into chapters. This will make the book an easy read for those who want to understand how to use UX programming theories in designing products.

The book starts with an introduction to the world of UX and the different theories that fall under it. This is followed by a chapter on product definition and how to identify your product's users. The next chapter is on product research, which covers different research methods, such as interviews, surveys, focus groups, and so on.

We hope that this book will help you create products that are easy to use and that meet the needs of your users.

Chapter 1

Introduction to UX and
What It Means for Your Brand

In this chapter, we will discuss:

- What is UX?

- What is UI?

- What is UX design?

- The goal of a good UX?

- How to create a good UX?

- Some principles of UX design

- A great UX is equal to a great UI

- How UX benefits your business

A Decent Design Is Indistinguishable
"Design isn't crafting a beautiful, textured button with
breath-taking animation. It's figuring out if there's a way to
get rid of the button altogether."
— Edward Tufte

So now that you have decided to pick up this book, let me promise you that you'll find awesome stuff here regarding product designing. And the fun part is that you can relate to this because you are a user yourself. You have probably used dozens of products, so in this book, you will have the opportunity to look at this from your point of view and learn what goes into making an awesome product or product design.

We'll begin by explaining what UX is. What's all the buzz about? And specifically, what is UX design? But since this book is about creating products using UX programming theories and methods, we will discuss every point in detail. So next time you search for an economically friendly dishwasher that does the work effectively, you'll know what went into making this innovative product.

What Is UX?

Simply put, UX refers to User Experience, how the user experiences the interaction with a different system, product, or service. Although it is mostly considered a computer term, UX and UI are above and beyond that. In all its beauty, UX is about how and what a user feels when interacting with different products. Many give it another name called human-computer interaction.

UX is also about a user's feelings towards a particular product or service. Were the features of the products approved by the user? What was the user's perception towards this product, and most importantly, how did a user feel and respond? You've heard about Customer satisfaction, no? Well, the better the UX, the greater the customer's satisfaction, which is what the brands strive for.

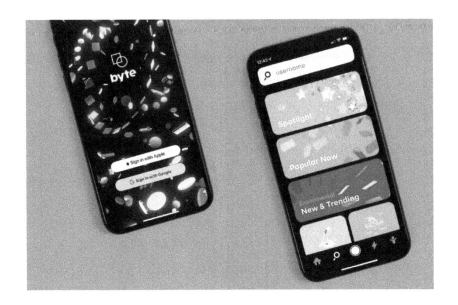

In other words, UX is about how a person feels when using a product or service. It's about the practical, experiential, emotional, and meaningful aspects of using a product or service. You can understand this point better by considering your own experiences as a user. How many times have you used a product or service that was difficult to use, confusing, or simply not enjoyable? How often have you found yourself thinking, *"I can't believe they made it so hard to do something so simple!"?*

Now, think about the products and services that you love using. What is it about them that makes them so enjoyable? Chances are that a big part of the answer is that they have great UX.

A great UX can turn first-time users into lifelong fans and turn satisfied customers into raving evangelists.

What Is UI?

The User interface (UI) is how a user interacts with a product or service. It includes the visuals and design of the product, as well as the overall experience of using the product.

What's the Difference between UX and UI?

UX is short for User Experience, which refers to how a user interacts with a product. UI is short for User Interface, which refers to the product's visuals.

A great UX design is important for creating a great product. The goal of UX design is to make the user's experience with the product as seamless and enjoyable as possible. This involves considering things like the user flow, navigation, and overall usability of the product.

A great UI design is also important for creating a great product. UI design aims to make the product visually appealing and easy to use. This involves choosing the right colors, fonts, and images for the product.

Both UX and UI are important for creating a great product. A product with a great UX but lousy UI will be difficult to use and may turn users away. A product with a great UI but lousy UX will be visually appealing but may be difficult to use and may also turn users away. The best products have both a great UX and a great UI. Let's take the example of a smartphone. A great UX is important for a smartphone because users need to be able to quickly and easily navigate the various features of the phone. A great UI is also important because users need to be visually stimulated to keep using the product.

How Do You Create a Great UX?

There is no one-size-fits-all answer to this question, as the best UX for a particular product or service will be unique to that product or service. However, some general principles can guide you in creating a great UX.

Creating a great UX is important because it can make the difference between a product that is used and loved and a product that is used and discarded. A great UX can turn first-time users into lifelong fans and turn satisfied customers into raving evangelists. It starts with understanding the needs of the user. It then takes into account all the factors that affect how a user experiences a product or service – from the practicalities of use to the emotions evoked by design.

What Is UX Design?

UX design is the process of creating products and services that are easy to use, enjoyable to use, and meet the user's needs. In other words, it is about making products and services people will love using. For example, let's say you want to keep your fitness in check and have downloaded an app for this purpose. As a UX rule of thumb, it should ask for your weight, your goal weight, height, and other necessary information on signing up so that it can give you a customized fitness plan keeping in view your personal data.

Principles of UX Design

If the UX is a planet, then its principle is the Sun. It cannot survive without its principles. To create the best possible user experience and the best possible product, every designer has to consider its principles. They are there for a reason. Think of them as a checklist that you just have to check off. These principles serve as a foundation and the building blocks of any great product. This chapter will briefly see these principles, each of which contributes to making an excellent product.

Some of the most important principles of UX design are:

1. Simplicity:

The best UX is often the simplest. When designing a product or service, always ask yourself: "Can we make this simpler?" A good rule of thumb is that if something can be done in fewer steps, it should be. A great example of this principle in action is the design of the Apple iPhone. The iPhone is widely considered to be one of the best examples of great UX, and one of the reasons for this is its simplicity.

2. Clarity:

A great UX is always clear. Users should never have to guess what they need to do in order to achieve their goals. All elements of the design should be easy to understand and use. For example, good labels, clear icons, and helpful error messages can improve a design's clarity.

3. Feedback:

A great UX provides users with feedback at every step of the way. This feedback can take many forms, such as error messages, progress indicators, auditory cues, haptic (touch) cues, or reassuring confirmation messages. This feedback lets users know they are on the right track and helps prevent errors. A great example of this principle in action is the use of notifications on a smartphone. Notifications give users timely information about what is happening with their apps, and they can take action if needed.

4. Ease of Use:

A great UX is easy to use. Users should be able to achieve their goals without any difficulty and the user interface should be intuitive and self-explanatory. Complex tasks should be broken down into smaller, simpler steps. All controls and features should be easy to access and use. For example, a user should easily be able to find the search box on a website, and the checkout process on an eCommerce site should be simple and straightforward.

5. Aesthetics:

A great UX is pleasing to look at and use. The overall look and feel of the product should be inviting and attractive. The user interface should be well-designed and visually appealing. For example, the use of whitespace, color, and typography should be carefully considered.

6. Consistency:

A great UX is consistent. All elements of the design should work together in a cohesive manner. The user interface should use similar conventions throughout so that users know what to expect. A great way to understand this principle is to think about the user experience of using a website. When you visit a new website, you expect certain things, such as the navigation being in the same place on every page. If a website deviates from these expectations, it can be confusing and frustrating to use.

7. Robustness:

A great UX is robust. It should be able to handle errors and unexpected inputs gracefully. Users should never lose data or progress because of a technical error. For example, a website should be able to handle unexpected traffic spikes without crashing.

8. Security:

A great UX is secure. Users should feel confident that their personal data is safe and secure. For example, a website should use SSL encryption to protect user data.

9. Privacy:

A great UX respects users' privacy. Users should be able to control how their personal data is used and shared. For example, a website should allow users to opt out of an account.

10. Accessibility:

A great UX is accessible. It should be usable by as many people as possible, regardless of ability or circumstance. For example, a website should be accessible to users with disabilities.

The American company Slack is a great case study of a product using UX methods.

Slack is a messaging app for teams that allows users to communicate easily with each other. It is simple to use, and its design is clean and uncluttered. It provides users with feedback at every step of the way and is easy to use. It is also visually appealing, with a consistent design throughout.

How Is a Great UX Related to a Great Product?

A great UX is essential for creating a great product. A great product must be easy to use and enjoyable to use. It must meet the needs of the user and provide them with a great experience. Therefore, a great UX is essential for creating a great product. Without a great UX, it is very difficult to create a great product.

When we talk about creating a great product, we must first consider the user, their needs, and what they need and want out of the product. The user is the most important part of any product, and so their needs must be at the forefront of the design. Only by understanding the user can we create a product that they will love to use.

A great UX is essential for meeting the needs of the user and providing them with a great experience. Consequently, a great UX is essential for creating a great product. Similarly, a great Product must consider that company's UX strategy. This means a company's design and implements its products must be based on a user-centric approach. This will result in products that are easy to use and enjoyable to use and also in products that users will want to keep using. In conclusion, we can say that a great UX is essential for creating a great product.

User experience (UX) design is about creating products that are easy to use and enjoyable. A great UX must be user-centered, meaning that the user's needs must be the primary focus of the design. It should also be simple and clear so that users can easily understand how to use it. Furthermore, it should provide feedback, so users know what actions they are taking and the results of their actions. Finally, a great UX must be visually appealing and consistent throughout.

Simplicity, clarity, feedback, and ease of use are essential elements of any great UX. However, the specific details of creating a great UX will be unique to each product or service. Therefore, it is important to consider the user's needs when designing any product or service. By doing so, you can create a great UX that people will love to use.

Now that we know that to build a product that your customers will actually love, you need to rethink your UX strategy. Let us now move on to how UX is beneficial for your business. For your understanding as a reader, I have provided some real-life scenarios that will help you understand the many advantages that a good UX provides.

How Does UX Benefit Your Business?

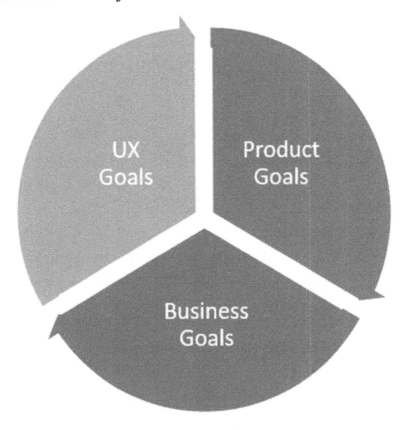

There are many benefits of incorporating UX into your business. Perhaps the most important benefit is that it can help you better understand your customers and their needs. By understanding your customers, you can design products that they will love to use. Additionally, UX can help to improve customer satisfaction and loyalty. Furthermore, it can help to increase sales and profits.

Finally, it can help to improve employee morale and productivity. There are many benefits of incorporating UX into your business.

In the year 2013, a study was done by Forrester on how UX affects business. The study found that companies who invested in UX saw an increase in sales, customer satisfaction, and market share. Additionally, these companies also had lower costs and higher employee morale. In conclusion, the study showed that investing in UX can positively impact business.

If you have used Duolingo, the popular translation app, you know how using UX features can enhance a product or service. Duolingo's design is based on user feedback; they use a simple and clear interface and provide feedback to users as they progress through the app. Plus, they provide their users with a seamless and enjoyable experience.

Another example of a company that saw improvement in its sales due to better UX and UI design is Airbnb.

Airbnb saw a 29.4% increase in its conversion rate after redesigning its website with a focus on UX and UI. They could improve their sales and conversions by providing their users with a better experience.

In conclusion, it is clear that incorporating UX into your product or service can have a positive impact on your business.

UX Goals = Product Goals = Business Goals

You can only create great products as a company if all three of these goals align together.

Creating a great product is not possible without a great UX. To create a great UX, you need to understand your users and their needs. Additionally, you need to create a product that is easy to use and enjoyable to use.

Knowing the Advantages of UX for Your Brand

Regardless of the type of organization, a strong and successful UX supports that business in numerous ways.

Firstly, understanding UX may aid in the efficient and reliable sale of more products and the portability of those items after clients have purchased them. By demonstrating to customers how well your business or organization serves its societies and staff through interesting corporate data, such as films on the support of the nearby charity, UX may also alter users' perceptions of your business or your brand. Less error-prone apps and interactions may save you money. By investing time upfront in creating user-friendly interactions, future revision is avoided.

If you're anything like the majority of us, you either run a company, work for one, plan to work for one, or are beginning one. Identifying your company objectives is the first step in figuring out how UX assists you. You really should be aware of your wider company goals; this book doesn't really assist you in defining them. Nevertheless, UX might influence some of your company's diversity and productivity.

The following objectives are usual for companies who create websites:

Keeping existing customers by giving them a helpful encounter and fostering their connection with the business (through customer support, news on their previous purchases, and value-added content that suits their current and future needs). Gaining new clients by employing a range of strategies, such as enhancing the user experience by including capabilities and details that rivals neglect to include, making it simple for clients to achieve their tasks and objectives, and successfully advertising the expertise in places like search engines.

Attempting to sell services or products by either up-selling or cross-selling equipment or objects to accompany the item. For instance, whenever a person acquires an outfit online, they are prompted to buy a pair of shoes and a belt to match the outfit. Giving updates about company operations, including a recent product announcement or leadership changes, business information, corporate communications information, details about the group's civic responsibility, etc., are all ways companies try to keep their customers up to date. There are many different methods to communicate this data, including articles, movies, and infographics (charts and graphics).

A strong user experience may provide the following advantages:

A rise in client satisfaction

Since UX is user-centered, following best practices for UX implies creating functionality, services, and information that are pertinent to consumers and their requirements. When users make a purchase, keeping them pleased requires paying attention to what matters to them. Customers will interact with a business repeatedly when they are happy with it. Additionally, he'll recommend it to their pals. This entails sharing a satisfying encounter on social media for numerous individuals. This principle is demonstrated by the example that follows.

A well-known adventure park in New York reportedly made an effort to lessen the inconvenience and agitation of standing in long queues. Jessica and her daughter are standing in a queue at an adventure park alongside around 125 other individuals when she witnesses this attempt personally. She comes upon notice with the words "Venture Off from a Long Queue: Seniors (check below), Teenagers (scan below), or Infants (scan below)" with a Barcode. She is encouraged to download the application after scanning it, allowing her kid to play with a game and a racetrack animation simultaneously. Although a 20-minute delay, the toddler is calmed. The program "detects" Jessica's position in the queue and adjusts the session to suit her surroundings: The game is completed when she gets to the forefront of the queue. She can post her reaction on Facebook and get a ticket for free cotton candy at the end of the queue with one more scanning. Jessica urges all of her female friends who have kids to visit the amusement parks. Usability testing, consumer surveys, and a strong UX strategy all contribute to the overall success of the situation.

Effect on good or service sales:

Since the user enjoys a straightforward, engaging, and instinctive purchasing experience overall, the overall revenue process increases when a UX is simple for a consumer or potential customer. But a strong UX has extra advantages.

A visitor is more inclined to trust a webpage if she believes it to be reliable and grasps her wants (and brand). A customer is more capable of completing a purchase or registering for an account when a business earns their confidence since the customer feels more confident disclosing personal details. Focus on the following instance: Natalie intends to get a fresh pair of sandals for a gala dinner. She observes that her preferred retail portal falls short of offering the design she wants to buy. After finishing a Google search, she is taken to a webpage with hundreds of footwear.

She observes that the website's shoe area includes a sizing chart, advice for different foot shapes (narrow versus broad soles), and a color-coordinating function. The website even suggests a bracelet and pendant as you're checking out! Natalie chooses to receive updates about upcoming deals as a result of her generally fulfilling experience. The webpage capability boosts sales performance while lowering the frequency of returns for poor fit.

Chapter 2

Understanding the Product Phases

In this chapter, we will discuss:

- The different product phases

- What is a product?

- Why do products fail?

- What is product design documentation?

- Why is having product design documentation important?

"Even the best designers produce successful products only if their designs solve the right problems. A wonderful interface to the wrong features will fail."
— *Jakob Nielsen*

This chapter will discuss the important phases a design team needs to carry out when developing a product. You may call it product development steps or phases.

Since we've moved to a more modern development process like *agile*, the focus is more on delivering products that are according to user needs at each step. We've given an overview of the product design and development phases. Every company and business has its own development phases, but the following are the commonly used phases.

When developing products or designing products, theory and practice have different approaches. Theories give us the framework to think about things, but they don't always reflect reality. The product development process is no different. There are many different ways to develop a product, and no one way is perfect. However, there are some common steps or phases that most companies go through when developing a product.

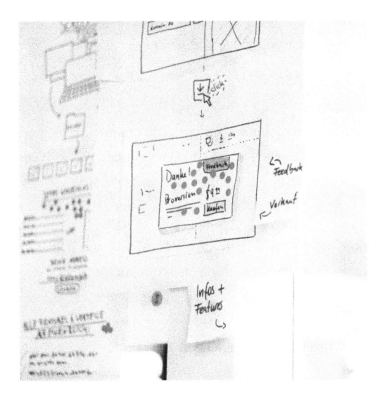

The following are the common steps or phases in product development:

- Research and planning

- Design

- Development

- Testing and launch

- Post-launch analysis/Iterate.

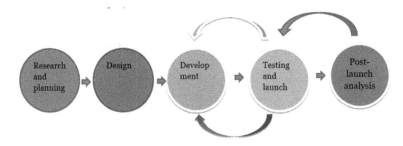

These are the common steps or phases in product development, but every company has its own way of doing things. When it comes to theoretical perspective, product documentation is an excellent example of product development is carried out.

Note: It is important to note here that this is not a linear model. There is a lot of back and forth between each of these steps. For example, you may develop a new feature and then realize that it needs to be designed differently. Or you may test the product and find some bugs that need to be fixed.

What Is Product Design Documentation?

Product design documentation is a written record of the product development process. It includes all of the information gathered during the research and planning phases and the design and development phases. Additionally, product design documentation should include information on testing and launch, as well as post-launch analysis.

Why Is Product Design Documentation Important?

Product design documentation is important because it provides a written record of the product development process. This documentation can be used to help improve future products. Additionally, product design documentation can be used to show how the product was developed and how it meets the needs of the user.

What Are Some Common Elements of Product Design Documentation?

Many different elements can be included in product design documentation. However, some of the most common elements include:

- A list of the goals and objectives for the product

- A description of the target audience

- A user flow diagram

- A wireframe or prototype of the product

- Information on how the product was tested

- A post-launch analysis.

These are just some of the common elements that can be included in product design documentation. Every company will have its own specific needs and requirements for its product design documentation.

We can all agree that UX is an important aspect to consider when designing products. It is important to align UX goals with business goals to create a product that is easy to use, enjoyable to use, and provides a seamless experience for the user. Additionally, it is important to understand the needs of the user to create a product that meets their needs. Finally, product design documentation is important because it provides a written record of the product development process. It can be used to help improve future products and show how the product was developed and how it meets the needs of the user

Going back to the phases of product development, let's discuss each phase step by step and look at how product development is carried out in each phase.

Research and Planning

In this phase, the team researches the problem or opportunity. They also define the goals and objectives of the product. This is a very important phase as it sets the foundation for the rest of the product development process. In this phase, documents such as stakeholder plans, project proposals, etc., are finalized with initial mock-up designs. Basically, this phase is all about brainstorming. The sophistication of the product, schedule, available resources, degree of background experience, and numerous other variables all affect this phase differently. Nevertheless, creating a marketplace and competitive studies and carrying out consumer surveys are generally smart ideas.

Design

In this phase, the team designs the solution to the problem that was identified in the previous phase. This includes creating wireframes, prototypes, and user flows. The documents in this phase are high-level and don't go into too much detail. For example, a wireframe is a low-fidelity representation of the product. Diagrams, schematic capture, demos, work diagrams, and design parameters are examples of common paperwork. For instance, user-personas developed in research and planning are included in mock-ups, mind maps, and simulations. These components then have an impact on advanced outputs, including concept art and thorough mock-ups.

Development:

In this phase, the team starts to build the product. This includes writing code, creating assets, and setting up infrastructure. The goal in this phase is to get a working product that users can use. For instance, when building a web application, the team would focus on getting the basic functionality working.

Testing:

The team tests the product in this phase to find any bugs or errors. This is done by both automated and manual testing. Additionally, user feedback is gathered in this phase to see if the product meets their needs. For instance, A/B testing is a type of testing that is used to compare two versions of a product.

Launch:

In this phase, the product is launched to the public and made available to users. This can be done through a beta release, public release, or private release. This includes creating marketing collateral, setting up analytics, and doing user research. Additionally, the team needs to plan for post-launch activities such as bug fixes. In this phase, the team monitors the product to see how it is being used and if there are any issues.

Iterate:

This is an ongoing phase where the team constantly makes changes and improvements to the product based on user feedback. This can be done through new features, bug fixes, or design changes. This is an ongoing process that happens throughout the life of the product.

A great example of this is the Facebook News Feed. It is constantly being iterated based on user feedback.

Utilizing data metrics and statistics, monitoring and refining the product while it is in operation leads to continuous, data-driven product upgrading.

These are some common product development phases. However, many books and authors have given different names to these phases and have divided these phases into more categories. For example, if you pick some other UX or product development book, you may see stages such as

- Product development
- Research
- Analysis
- Designing
- Implementation
- Launching
- Measuring and iterating

What Are Some Common Challenges That Teams Face When Developing a Product?

One of the biggest challenges teams face when developing a product is scope creep. This is when the scope of the project starts to increase without any increase in the budget or timeline. This can happen for a number of reasons, such as new features being added, changes in the design, or unforeseen problems. Another challenge is managing stakeholders. This is when multiple people need to be kept up-to-date on the project's progress and decision-making. This can be difficult as everyone has their own opinions and needs. Managing dependencies is another great challenge. This is when there are other teams or products that the current team is reliant on. This can cause delays if there are any problems with the other team or product

The steps in product development are important, but they don't always reflect reality. For example, the design and development phases may overlap in a real-world scenario. Additionally, the product may have multiple iterations before it is launched.

Another common challenge is getting user feedback. This can be difficult to do if the product is not yet launched. One way to get user feedback is through beta testing. This is when a group of users test the product and provide feedback. Another way to get user feedback is through user research. This is when the team interviews users and asks them about their needs and wants.

Socially Defined Strategies

After seeing how each phase is interconnected, let's examine some useful guidelines for advancing the idea through each step. We'll go over ways to implement design iterations so that the procedure changes over time rather than being initially established. Design sprints are 1-3 week iterations that concentrate on resolving particular product and designing concerns, much like their Agile software equivalent. Alok Jain, UX Manager at 3Pillar, identifies communication, decreased transfer barriers, and team concentration as the three essential components of design sprints. In a word, your documentation is a team project that must constantly keep the user in mind. You develop energy and reduce waste since you transition swiftly between each phase. The fact that you're working on lesser issues enables more experimentation and risk-taking, which is more significant.

Chapter 3

The Product Phases

In this chapter, we will discuss:

- Each phase of the product in detail

- What is a product by definition?

- What is product designing?

- What is product development?

- What happens when you test a product?

- The ways through which developers conduct product testing

- The launch phase.

"We try to solve very complicated problems without letting people know how complicated the problem was."
— Jony Ive

In the previous chapter, we looked at each phase of the product from a bird's eye view without going into much detail. Trust us: a lot happens behind the curtains. The smartphone in your hand isn't there just because the owner or stakeholders thought of making a 6-

inch smartphone with a sleek, thin body. It's there because it answers users like your needs. It's there because a lot of research has gone into finding what features users would like to see in a phone.

We will now look deeply into each phase and see what happens in each stage of product development. What documents are made? What do the design and the development teams do, and what techniques are used in each stage?

Understanding the Product

It is very important to take the time to understand the product that you are working on. This means understanding the features of the product and how users will interact with it. It is also important to understand the business goals of the product. All of this information will help inform your design decisions. You should ask yourself questions like:

- *What are the features of the product?*

- *How will users interact with the product?*

- *What are the business goals of the product?*

According to UX designer Jesse James Garrett, there are five planes of a product:

- *The Strategic Plane:* This is where you define the goals of the product.

- ***The Scope Plane:*** This is where you define the features of the product.

- ***The Structure Plane:*** This is where you determine how the product will be organized.

- ***The Skeleton Plane:*** This is where you create the interface for the product.

- ***The Surface Plane:*** This is where you determine the look and feel of the product

While trying to figure out the product, features, and user market, you should also consider what kind of team you will need to develop the product. Do you need a team of developers? A team of designers? A team of both? Putting together the right team to create a great product is important.

A great case study of product definition is the development of the iPhone. The original iPhone was released in 2007 and revolutionized the smartphone industry. The iPhone was designed to be an easy-to-use device that combined the best features of a cell phone and a desktop computer. The iPhone was also designed to be visually appealing with its sleek design and large touchscreen display.

Smashing magazine has a great article that outlines the process the Apple team went through to develop the iPhone. The article is called "The Design and Development Process of the Original iPhone." We recommend you look it up and give it a read.

Now the question arises: What do design teams do in the *product-defining* stage? They just think about the product, its features, and how users will interact with it, or is there something more to it?

Yes, design teams do more than just think about the product, its features, and user interactions. They also think about the *product's business goals* and how best to organize the product. All of this information helps inform their design decisions. *Stakeholders* will also be involved in this stage to help provide input on the product goals and features.

Brainstorming is one of the most important aspects of this stage. It is important to have various ideas to choose from to create the best product possible. Brainstorming can be done in a variety of ways, but some common methods are:

- *Individual brainstorming*: This is where each team member comes up with ideas on their own.

- *Group brainstorming*: This is where the team comes together and brainstorms ideas as a group.

- *Online brainstorming*: This is where the team uses an online tool to brainstorm ideas.

Some popular online brainstorming tools are:

- Google Docs

- Mural

- Stormboard

Once you have a list of ideas, it is important to start *narrowing* them down. This can be done by *ranking* the ideas in order of importance or by using a *criteria matrix*. A criteria matrix is a table that is used to evaluate ideas based on a set of criteria. This is a great way to compare and contrast ideas objectively.

Once you have narrowed down the list of ideas, it is time to start *creating prototypes*. A *prototype* is a mock-up of the product that can be used to test out ideas. Prototypes can be low-fidelity (simple and quick to create) or high-fidelity (more detailed and takes longer to create). It is important to create prototypes early on in the process so that you can get feedback from users and make changes as needed.

Okay, but what about documentation? Surely the teams will need and will be creating documents in each phase? What are some of the documents created in this product-defining stage?

Yes! As in any business, documents are an integral part of the whole process. While most of the documents are produced early in the beginning phase, they keep on updating in each phase.

A few different types of documentation are created in this stage.

- The first is the product requirements document. This document outlines the goals, features, and requirements of the product.

- The second is the functional specification document. This document outlines how the product will work and includes detailed user interface and functionality information.

- The third is the design specification document. This document outlines the product's visual design and includes information on color, typography, layout, and iconography.

Let me tell you guys a true story of how a company defined its product using UX principles. The company in question is Apple, and the product is, of course, the iPhone. The original iPhone was released in 2007 and was a game changer. It was the first smartphone that really caught the public's attention, and it set the standard for what a smartphone should be.

When the iPhone was first developed, the Apple team knew they wanted to create a truly unique and groundbreaking product. To do this, they had to think outside the box and come up with something that had never been done before.

They started by looking at what other smartphones on the market were doing, and they quickly realized that most of them were focused on business users. They decided to take a different approach and focus on the average consumer's needs.

The management knew that the user experience would be a key part of the product. They wanted to ensure that the iPhone was easy to use and had an intuitive interface. In order to achieve this, they developed a new type of user interface that was based on direct manipulation. This meant that users could interact with the phone by using their fingers to manipulate the on-screen elements directly.

Another important aspect of the product was the design. The team at Apple wanted to create a product that was not only functional but also stylish. They wanted the iPhone to be something that people would want to show off. To achieve this, they worked with a team of industrial designers to come up with a sleek and elegant design.

The result was a product that was unlike anything that had been on the market before. The iPhone was a success, and it changed how people thought about smartphones.

Once you have the basic ideas, it is time you do your user research. You have to consult your *potential users* and try to gauge their reactions. Try getting as much feedback as possible at this stage because it will help you make informed decisions about the product.

You can carry out *interviews*, *surveys*, *focus groups,* or even just have casual conversations with people to get their feedback.

You can start refining the product ideas based on the feedback you receive. You can also start creating *personas* at this stage. Personas are fictional characters that represent your *target users*. Creating personas is a great way to help the team understand the needs and wants of your target users.

When defining the product, always keep the user in mind. The product should be designed in such a way that it is easy for the user to accomplish their goals.

Designing the Product

After you have done all the groundwork regarding your *product's definition*, you can now move on to designing the product. The first step in this process is to create a *wireframe*. A wireframe is a low-fidelity mock-up of the product that shows the content's layout and the user interface's general structure. Wireframes are used to get an early idea of what the product will look like and test out different design ideas.

Once you have a wireframe that you are happy with, you can move on to creating high-fidelity mock-ups. These are more detailed versions of the wireframes, including actual content and visual elements. High-fidelity mock-ups are used to get a better idea of the overall look and feel of the product.

Prototypes are also created at this stage. ***Prototypes*** are working versions of the product that allow users to interact with it and test out its features. The documentation in this stage will also be more detailed, including things like the user flows and the interaction design. You have to consider some points while you are designing the product. This includes Sketches, which will help you to materialize your ideas quickly, user flows to help you understand how the user will interact with your product, and interaction design to define how the user will interact with the individual elements on the screen.

How does the design teamwork in this phase? Well, the design team will take care of all the graphical aspects of the product, and they will also work on the user experience. According to Apple's ***Human Interface Guidelines***, three main principles should be kept in mind when designing the user experience:

1. The User

The development process should always be focused on the needs of the user. The user should be able to accomplish their goals simply and efficiently.

2. Direct Manipulation

The user interface should be based on direct manipulation. This means that users should be able to interact with the on-screen elements using their fingers.

3. Feedback

The user should receive feedback in a timely manner. This feedback can be in the form of visual, auditory, or haptic cues. According to UXPin, the following steps should be followed when designing the user experience:

1. Define the problem

2. Come up with solutions

3. Choose the best solution

4. Implement the solution

5. Test the solution

6. Iterate

Yelp is a company that takes product design seriously and makes changes to its product based on user feedback.

1. They talked to their users and found out what they wanted

2. They made changes to their product based on that feedback

3. They tested the new product with users

4. They made further changes based on the feedback they received

5. They launched the new product

This is a great example of how important it is to involve users in the product design process. By talking to users and getting their feedback, Yelp was able to make changes that greatly improved its product.

It is important to note here that designing a product works in iterations. This means the product will go through several rounds of design, development, and testing before it is launched. You may have to change your design strategy based on feedback from users or from your development team. Don't be afraid to experiment and try new things! The most important thing is to create a product that the user will love.

So what are some techniques through which the design team designs products?

Apart from prototypes and mock-ups, **Brainstorming** and **user testing** are two other popular and effective methods that the design team uses to come up with the best product designs.

1. **Brainstorming:** This is a technique where the design team comes up with ideas by brainstorming. This can be done either in person or online.

2. **User testing**: User testing is a great way to get feedback on prototypes. User testing can be done using paper or digital prototypes.

We recommend using a simple pen and paper to design your product. This will help you to communicate your ideas clearly to the team. You can also use digital tools such as Sketch or Photoshop to create

Developing the Product

After the product designing phase, we can actually sit and start *developing* the product. Here we will be using different programming languages to write codes that will make the product work according to our design. But as you know, not everyone is a developer, and not everyone can understand how these codes work. However, there are some basic concepts of UX programming that everyone should know to create great products. In this book, we will be discussing some of the most important concepts of UX programming. By the end of this book, you will have a better understanding of how these concepts can be used to create amazing products.

Back to the development stage, the *technicalities* of the product will be handled in this stage. This is where all the behind-the-scenes work happens to make the product look amazing and function perfectly. The design team will work closely with the development team to ensure everything goes smoothly and according to plan. The main deliverables of this stage are:

- The product prototype
- The coding of the product
- The testing of the product
- The launch of the product

This is where all the hard work pays off, and we see our product come to life. All our efforts in the previous stages will now be put to the test. We will see if our product is able to meet all the user requirements and deliver on our promises. If everything goes well, we will be able to launch our product successfully and see it being used by people worldwide.

You may have to go back to the previous stages to make changes and improve the product before launching it. But once it is out there, it is all worth it. For example, when you're developing the product, you may realize there is some sort of a minor flaw in the design that you didn't catch before. But that's okay because you can fix it in the development stage and ensure that it doesn't happen again. As we said before, this is not a linear model. Rather the product development works in iterations.

The famous UX manager and Agile lead, Jeff Sutherland, says, *"agile is all about constant learning and improvement."*

So even if you make a mistake, you can learn from it and become better for it. Even though in the previous stages you have a roadmap that you need to follow, in this stage, you need to be flexible and adaptable. Things will come up that you didn't expect, and you will have to find a way to solve them. This is where your creativity and problem-solving skills will be put to the test.

What models or techniques can you use to ensure your product is developed effectively? Let's take a look at some of them.

1. The Waterfall Model:

This is a very traditional approach to product development. In this model, each stage is completed one after the other in a linear fashion. So first, the requirements are gathered, then the design is created, followed by the development, testing, and finally launch. The main advantage of this model is that it is straightforward to understand. The main disadvantage is that it is very inflexible and does not allow for many changes once a stage has started. So if you realize that there is a problem in the design, you will have to wait until the development stage to fix it, which can be very costly and time-consuming.

2. The Agile Model:

This is a more modern and flexible approach to product development. In this model, the different stages are completed in iterations or sprints. So the requirements gathering, design, development, testing, and launch are all completed in small chunks. This allows for much more flexibility and allows you to make changes along the way. The main advantage of this model is that it is very adaptable and can easily accommodate changes. The main disadvantage is that it can be very chaotic and disorganized if not managed properly.

3. The Spiral Model:

This is a combination of the waterfall and agile models. In this model, the product is developed in a spiral shape. So first, a small part of the product is developed, and then it is tested. Based on the feedback from the testing, the next part of the product is developed, and so on. This allows for a lot of flexibility and also allows you to make changes along the way. The main advantage of this model is that it is very adaptable and can easily accommodate changes. The main disadvantage is that it can be very chaotic and disorganized if not managed properly.

4. The Lean Model:

This is a model that focuses on efficiency and waste reduction. In this model, the development team only works on the essential parts of the product. This allows them to focus on the most important things and avoid wasting time on unnecessary features. The main advantage of this model is that it is very efficient and can help you save a lot of time and money. The main disadvantage is that it can be very inflexible and may not allow for much creativity.

5. The Kanban Model:

This is a model that focuses on continuous improvement. In this model, the development team is divided into small teams, and each team works on a specific part of the product. The team then tracks their progress and makes changes based on feedback. This allows for a lot of flexibility and also allows you to make changes along the way.

So these are some of the most popular product development models. Which one you choose will depend on your specific needs and requirements. But whichever model you choose, ensure that you understand it well and have a clear plan for how to implement it. Otherwise, you may end up wasting a lot of time and money.

An important thing to note here is that when developing products, you need to pay special attention to the features of your products. And there is a model that helps development teams in this regard, and that model is called the *"Kano Model."*

The Kano model is very popular among product development teams, helping them prioritize the features of their products. And this is done by dividing the features into three categories.

- The first category is called *"must-have"* features. These are the features that are absolutely essential for the product to be successful. For example, if you are developing a messaging app, the must-have features include sending and receiving messages, creating groups, and so on.

- The second category is called *"nice-to-have"* features. These features would be nice to have but are not absolutely essential. For example, if you are developing a messaging app, the nice-to-have features would include video calling, voice calling, and so on.

- The third category is called *"do-not-want"* features. These are the features that you do not want in your product. For example, if you are developing a messaging app, the do-not-want features include spam filters, ads, and so on.

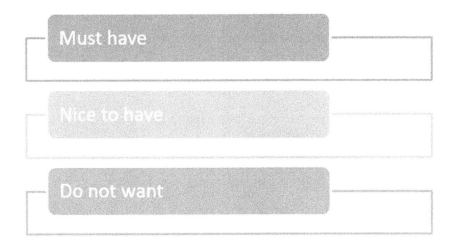

These are the three categories of features. And development teams need to prioritize the features based on these categories.

The main advantage of using the Kano model is that it helps development teams focus on their products' essential features. And this helps them to save a lot of time and money.

The main disadvantage of using the Kano model is that it can be very inflexible. And this is because the development team needs to strictly follow the categories. Otherwise, they may end up adding features that are not essential, and this can lead to a lot of wasted time and money.

Once you have listed your nice-to-have or must-have features, you can prioritize them by using a ***prioritization matrix or table***.

The prioritization matrix is a very popular tool among product development teams. This matrix helps development teams to prioritize the features of their products. Can you give them a score from 1-10 and plot them on the map? The advantage of doing this is that it gives you a basic idea or can serve as a map to navigate through the important and less important features you want in your product.

Netflix is a very popular streaming service. And they use a model called "Churn Prediction" to find the best features for their products. What they do is they take a sample of their customer base, and then they use machine learning to predict which features are most likely to cause customers to cancel their subscriptions. They then use this information to prioritize the features of their products. And as a result, they have been able to create a very successful product.

Similarly, Apple uses a model called "*Apple new product process*" to find the best features for its products.

They use a 6-step process to assess, ideate, prototype, test, and select the features for their products.

- The first step is to assess the need for a new product. They do this by looking at the market trends and the competition.

- The second step is to ideate or comes up with new ideas for their products. They do this by brainstorming with their team and looking at customer feedback.

- The third step is to prototype or creates a mock-up of their product. They do this so that they can test their product and get feedback from users.

- The fourth step is to test their product. They do this by releasing a beta version of their product and collecting feedback from users.

- The fifth step is to select the features of their product. They do this by looking at the feedback they receive from users and selecting the features that are most important to them.

- The sixth and final step is to launch their product.

As you can see, both Netflix and Apple use models to find the best features for their products. And these models have helped them to create very successful products.

In terms of the advantages these models provide, they surely provide the best features for your product as they help you save time, they help you to prioritize the features of your product, and they help you to create a successful product.

Testing the Product

After the development phase, it is important to test the product to make sure that it is working correctly. This can be done with beta testers or by releasing a limited version of the product. In this phase, it is also important to get feedback from users so that you can improve the product before the official launch. How is product testing done?

There are two main ways to test a product: beta testing and A/B testing.

1. Beta testing is when a product is released to a small group of users so that they can test it and provide feedback. This is usually done before the official launch of the product. For example, Facebook released a beta version of its News Feed feature to a small group of users before rolling it out to all users.

2. A/B testing is when a product is released to two groups of users, and each group sees a different version of the product. This is done so that you can see which version of the product is more successful. For example, Google has done A/B testing with its search results and has found that adding an extra line of results can lead to a 5% increase in clicks.

Regarding the documents that are made in this phase, some of them are given below. As always, they are updated as new signs of progress are carried out.

Functional specification: This document outlines the product's features and how they should work. An example of this is the requirements document that is made in the planning phase.

Technical specification: This document outlines how the product will be built and what technologies will be used. The development team usually does this. For example, if you are making a website, the technical specification would outline what programming languages will be used and how the website will be hosted.

Product roadmap: This document outlines the plans for the product and shows how it will be developed over time. The product manager usually does this. For instance, the product roadmap for Facebook's News Feed feature would show when it will be released to all users and what improvements will be made over time. Remember, these documents are also made early in the brainstorming phase but are updated in every phase.

Design specification: This document outlines the design of the product and how it will look. For example, if you are making a website, the design specification would show what the website will look like and how users will interact with it. The design team usually makes this.

User experience specification: This document outlines the product's user experience and how it will feel. For example, if you are making a website, the user experience specification would show what the website will feel like and how users will interact with it. The user experience team usually makes this.

Hewlett-Packard (HP) is a technology company that designs and manufactures computer hardware, software, and services. The company was founded in 1939 by Bill HP and Dave Packard, and it is headquartered in Palo Alto, California.

HP conducts user testing to improve the usability of its products. The company has a dedicated usability lab where users can test new products and give feedback. HP also uses focus groups to get feedback from users. In addition, the company uses online surveys and customer feedback to gather data about how users interact with its products.

If a product fails in the *testing stage*, it usually means that there are some major flaws with the design or functionality of the product. This can be caused by several things, such as an incorrect assumption about how users would interact with the product or a problem with the technology used to build the product. If a product fails in the testing stage, it is usually necessary to go back to the drawing board and redesign the product.

What is the most important thing to remember about this phase?

The most important thing to remember about this phase is that it is important to get feedback from users early and often. This will help you to identify any problems with the product before it is released to the public.

Launching the Product

After the product is built and tested, it is time to launch it. This is when the product is made available to users. For example, if you are making a website, the launch phase would involve putting the website live on the internet. In this phase, it is important to monitor the performance of the product and collect feedback from users. The development and design team will use this feedback to improve the product over time. However, in order to make the products successful, the team must ensure that some goals are met, which are:

- *Creating a great user experience*: This is the team's most important goal as it ensures that users will enjoy using the product and will keep coming back.

- *Creating a usable product*: The product must be easy to use and navigate. Users should be able to find what they are looking for easily.

- *Meeting the user's needs: The product must meet the user's needs*. For example, if you are making a website for a business, the website must be able to sell products or services.

- *Creating a scalable product*: The product must be able to handle an increase in users. For example, if you are making a website, the website must be able to handle an increase in traffic.

The development and design team must also ensure that the product is launched on time and within budget. This can be a challenge, but it is important to remember that the product will not be successful if it is not launched on time. Furthermore, KPIs and other metrics must be monitored to assess the product's success. Metrics tools such as Google Analytics can be used to track KPIs. Many companies have their launch goals set before the product is even built. This ensures that everyone is working towards the same goal and that the product is launched on time. Companies such as Facebook and Google often launch products in beta versions to get feedback from users before launching the product to the public.

The product launch is a critical moment for the development and design team. If the product is not launched on time, it can negatively impact the company. Furthermore, if the product is not well-designed, it will not be successful. Therefore, it is important to remember the development phase's importance and avoid any potential pitfalls.

Some methods to launch a product are:

1. ***Creating a beta version:*** This is when a product is released to a small group of users to get feedback before launching it to the public.

2. ***Launching in stages:*** This is when a product is launched in different stages to different groups of users. For example, a product may be launched first to early adopters and then to the general public.

3. ***Doing a soft launch:*** This is when a product is launched in a limited way to test the market. For example, a product may be launched in a small geographical area before the rest of the world.

4. ***Giving away freebies:*** This is when companies give away free products or services to promote a product. For example, a company may give away free samples of a new product.

Twitter is a good example of a company that used beta testing to launch its product. The product was launched in 2006 in the beta version. This allowed the company to get feedback from users and make improvements before launching the product to the public. The product was officially launched in 2007 and has been a success since then. In 2015, the company had over 300 million active users.

According to a report by TheNextWeb, Apple used a soft launch to launch its Apple Music service in 2015. The service was first launched in 100 countries. This allowed the company to test the market and make sure that the service was working properly before launching it in more countries. The service was officially launched in all countries in 2016.

What are some of the documents created during the launch phase of the product?

Some of the documents that are created during the launch phase of a product are:

1. *A marketing plan:* This document outlines how the product will be marketed to users. For instance, it will include information on how to generate awareness and interest in the product.

2. *A user manual:* This document provides instructions on how to use the product. For example, it will explain how to set up the product and how to use its features.

3. *A technical manual:* This document provides information on the technical aspects of the product. For example, it will explain how the product works and how to troubleshoot any problems that may occur.

4. *A support plan:* This document outlines how the product will be supported after it is launched. For example, it will explain what type of customer support will be available and how to contact the support team.

5. *A sales plan:* This document outlines how the product will be sold to users. For example, it will include information on generating interest in the product and converting users into paying customers.

6. *A pricing strategy:* This document outlines how the product will be priced. For example, it will include information on whether the product will be free or paid, and how much it will cost if it is paid.

7. *A business plan:* This document outlines the business goals and objectives for the product. For example, it will explain how the product will generate revenue and how it will be profitable.

Chapter 4

Defining a Product

In this chapter, we will discuss:

- What is a product by definition?

- Why designing a product is important

- What are the components of a product

- Some methods through which you can find and define your product

- Product Definition Process

> *Good UX is good business.*
> *– Andrew Kucheriavy*

The first step in the product definition process is, of course, to **define the product**. This seems like a simple task, but it's actually quite difficult. There are a lot of factors to consider, and if even one is missed, it can throw off the entire design.

To help you out, we've put together a list of questions you should ask when defining a product:

1. Who is the user?

2. What does the user want to accomplish?

3. What are the user's goals?

4. What are the user's pain points?

5. What is the competitive landscape?

6. What are the company's goals?

7. What are the technical constraints?

8. What is the timeline?

9. What is the budget?

10. What is the product's name?

11. What is the product's tagline?

12. What is the product's elevator pitch?

The *Product Definition* phase is critical to the success of any product. By taking the time to answer these questions accurately, you'll be setting yourself up for a smooth design process.

Without a solid product definition, it's all too easy to get lost in the design process. So take your time, ask lots of questions, and ensure you clearly understand the product before moving on to design. You should also involve as many stakeholders as possible in the product definition process. The more people you have involved, the better the final product will be.

During this stage, it's also important to consider the user experience (UX). How will users interact with the product? What features do they need? What kind of interface will they be using? These are all important questions that should be answered during the product definition phase. Activities such as brainstorming and user research will be particularly helpful during this stage. All of these activities result in a better understanding of the user, which in turn results in a better product.

Why Defining a Product Is Important

The first step in creating any product is defining what the product is. This may seem like a simple task, but it's actually quite difficult and important. When defining a product, there are many factors to consider, and if even one of them is missed, it can throw off the entire design process.

Creating a clear definition for a product is important because it provides a foundation for the rest of the product development process. It's much easier to design and build a product when everyone involved knows exactly what the product is supposed to be.

The three important components that you need to keep in mind when defining a product are Feasibility, Viability, and Usability.

What Is a Feasibility Study?

A feasibility study evaluates whether or not a proposed product can be successful. It's important to conduct a feasibility study early on in the product development process because it can save a lot of time and money if the product is not feasible. There are a few different things that you need to consider when conducting a feasibility study:

1. *The market*: Is there a market for the product? Will people actually buy it?

2. *The technology*: Can the product be built using available technology?

3. *The resources*: Do we have the time, money, and manpower to build the product?

What Is a Viability Study?

A viability study evaluates whether or not a proposed product can be profitable. It's important to conduct a viability study early on in the product development process because it can save a lot of time and money if the product is not viable. *There are a few different things that you need to consider when conducting a viability study:*

1. **The market**: Is there a market for the product? Will people actually buy it?

2. **The competition**: Is there already a product like this on the market? If so, can we compete?

3. **The cost**: How much will it cost to produce the product? Can we sell it for a profit?

What Is Usability Testing?

Usability testing is a process of testing how easy it is to use a product. It's important to conduct usability testing early on in the product development process because it can save a lot of time and money if the product is not usable. There are a few different things that you need to consider when conducting usability testing:

1. The user: Who will be using the product? Do they have the skills necessary to use it?

2. The task: What task are we testing? Is it a common task that users will need to do?

3. The environment: In what environment will the user be using the product? Is it a familiar environment?

4. The results: Did the user complete the task? If not, why not?

Conducting a feasibility study, viability study, and usability testing are all important components of the product definition phase. They will help you to understand the user and the market better, and ultimately result in a better product.

Let us take a case study here and explain all of these three points in detail:

You are designing a new e-commerce app, and you want to make sure that it is successful. To do this, you need to make sure that it is feasible to build, that there is a market for it, and that it is usable.

The first thing you need to do is conduct a feasibility study. This will help you determine if the app can be built using available technology and if you have the time, money, and manpower to do so. If you find that the app is not feasible, you may need to change your plans.

The second thing you need to do is conduct a viability study. This will help you determine if there is a market for the app and if you can compete with other apps on the market. If you find that the app is not viable, you may need to change your plans.

The third thing you need to do is conduct usability testing. This will help you determine if users can actually use the app and if they can complete common tasks. If you find that the app is not usable, you may need to change your plans.

By conducting these three studies, you will be able to understand the user market better. These studies will ultimately result in a better product.

According to Luke Wroblewski, "The first step in any design process is understanding the problem you're trying to solve. This step is especially critical on web projects because the web is rife with problems that have already been solved."

This couldn't be truer when applied to product design. To design a great product, you first need to understand the problem that you're trying to solve. This means conducting research, talking to users, and understanding the market. Only then can you start to design a solution that meets the needs of the user and the market.

According to Mark Curphey, the founder of the software development consultancy, "The product definition stage is the most important part of the product development process. It's where you define the product and its features. It's also where you determine if the product is feasible and viable. If the product is not feasible or viable, it's important to go back to the drawing board and make changes before moving on to the next stage."

Pro tip: Make sure to involve all stakeholders in the product definition process. This includes the product owner, the development team, and other members of the organization who will be affected by the product. By involving all stakeholders, you can ensure that everyone is on the same page and that the product definition is accurate.

How Do You Define a Product?

There are many different ways to define a product. The most important thing is to ensure that all key stakeholders are involved in the process. This includes the product owner, developers, designers, and users.

Some common methods for defining a product include:

1. Brainstorming

2. User research

3. Competitive analysis

4. Creating personas

5. Writing user stories

6. Developing use cases

7. Creating a product roadmap

Let us know and take a closer look at the activities that need to be carried out when defining a product

The first activity is to create a stakeholder map.

A stakeholder map is a simple way to visualize all people with a vested interest in the product. This includes the development team, the design team, management, and of course, the users. By creating a stakeholder map, you can ensure that everyone is aware of the product and its role in its development. For example, the development team may not be aware of the product's features, while the design team may not be aware of the product's roadmap.

The second activity is to create user personas.

User personas are fictional characters that represent the different types of users that will be using the product. By creating user personas, you can better understand the needs of the user and design the product accordingly. For example, if you're designing a social media app, you might create a user persona who is a teenage girl. This persona would have different needs than a user persona who is a middle-aged man.

The third activity is to create a product roadmap.

A product roadmap is a document that outlines the development of the product. It includes the features that will be included in each release and the timeline for each release. By creating a product roadmap, you can ensure that the product is developed promptly and efficiently.

Taking a closer look at the different aspects of a product, it is imperative that we also take into account the levels of a product, which are:

- *Core product*

- *Actual product*

- *Augmented product*

The core product is the basic need that the customer is trying to satisfy, such as the need for transportation, shelter, or information. For instance, when we buy a car, we don't just pay for the vehicle itself but also for the features that come with it, such as the engine, the wheels, and the brakes.

The actual product is what the customer receives when they purchase the product. This includes all of the features and benefits included in the product. For example, when we buy a car, we receive not only the vehicle itself but also the warranty, the customer service, and the financing options that come with it.

The augmented product is the total package the customer receives when purchasing it. This includes the core product, the actual product, and all of the additional services that are included in the purchase. For example, when we buy a car, We also receive the option to add on additional features, such as a GPS system or an extended warranty.

Let us take a case study to help you understand this concept:

Suppose you want to buy a new car. The company offers you the following product levels:

- **Core product**: The basic need that the customer is trying to satisfy, such as the need for transportation.

- **Actual product**: The car itself, plus the warranty, the customer service, and the financing options that come with it.

- **Augmented product**: The total package the customer receives when purchasing the car. This includes the core product, the actual product, and all of the additional services that are included in the purchase.

Let us take the example of a mobile app that allows users to order food from restaurants.

The core product would be the ordering system itself. The augmented product would be the GPS system that allows users to find restaurants near them, and the actual product would be the mobile app.

Since Products are so diversified, it is important to understand how they are created. In order to create a product, businesses go through what is called the Product Definition Process. This process aims to develop a clear and concise definition of the product before it is developed. This ensures that all stakeholders are on the same page and that the product meets the needs of the customer.

The Product Definition Process consists of four steps:

1. Define the customer's need

2. Develop the product concept

3. Create the product specification

4. Test the product concept

Let's take a closer look at each of these steps:

1. **Define the customer need:** The first step in the Product Definition Process is to define the customer need. This involves understanding the problem the customer is trying to solve and what they are looking for in a product.

2. **Develop the product concept**: The next step is to develop the product concept. This involves creating a vision for the product and determining what the product will do for the customer.

3. **Create the product specification**: The third step is to create the product specification. This document outlines the details of the product, such as the materials to be used, the manufacturing process, the packaging, and the labeling.

4. **Test the product concept**: The fourth and final step is to test the product concept. This involves testing the product with potential customers to get feedback and ensure that the product meets their needs.

Let us take the help of an example to help you understand these processes:

You are the owner of a small business that sells car parts. Recently, you have been noticing that your sales have been declining. After doing some research, you have determined that the reason for this decline is that your customers cannot find the car parts they need. They are also telling you that they are not happy with the quality of your car parts.

To solve this problem, you need to go through the Product Definition Process.

The first step is to define the customer's need. In this case, the customer's need is for car parts that are easier to find and of a higher quality.

The next step is to develop the product concept. For this, you need to determine what your product will do for the customer. In this case, you decide that your product will be a website that allows customers to find the car parts they need easily.

The third step is to create the product specification. This document will outline the details of your product, such as the materials that will be used, the manufacturing process, the packaging, and the labeling.

The fourth and final step is to test the product concept. This involves testing the product with potential customers to get feedback and ensure that the product meets their needs. After getting feedback from customers, you make some changes to your product and launch it on your website.

Now that you understand the Product Definition Process, you can use it to develop your own products. By following this process, you can ensure that your products meet your customers' needs and that they are satisfied with the final product. Some more points need to be considered when defining your product. A lot happens behind the curtains. Let us take you through these points step by step

Kick-Off Meeting

The product definition process begins with a kick-off meeting. This is a meeting between the product stakeholders, such as the product owner, the development team, and the marketing team. The purpose of this meeting is to discuss the product and decide on the project's overall direction. Some guidelines should be followed when commencing a kick-off meeting.

The first step is actually preparing for the kick-off meeting. This means that the stakeholders should come to the meeting with an open mind and be ready to discuss the product. It is also important to have a clear agenda for the meeting. This will help keep the meeting on track and ensure that all important topics are covered. Think of the kick-off meeting as a welcoming ceremony for the product. It is of great importance that you set the right expectations so that everyone is on the same page from the beginning. The kick-off meeting should discuss everything from the project budget to deadlines. *According to Atlassian*, a good kick-off meeting should answer the following questions:

1.   What is the product?

2. Who are the stakeholders?

3. What are the objectives of the product?

4. What are the user needs?

5. Who are the users?

6. What are the user stories?

7. What is the product backlog?

8. What are the sprints?

9. What is the definition of done?

10. How will progress be measured?

As part of the Kick-off meeting, you should:

Define Product Goals:

The first step in starting a product is to define the goals of the product. These goals should be specific, measurable, attainable, relevant, and time-bound. In other words, they should answer the question of what the product is supposed to do.

Some examples of product goals are:

- Increase sales by 10% in the next quarter

- Improve customer satisfaction scores by 5%

- Reduce customer churn by 2%

All stakeholders should discuss and agree upon these goals before moving on to the next step.

Defining Your Vision:

The next step is to define your vision. This means that you need to determine what you want the product to achieve. For example, if you are creating a website, your vision might be to increase web traffic by 20%. Once you have defined your vision, you can move on to the next step. You can't just make or design products. You have to define what you want it to achieve and how. You need to have an image, a vision in your mind of what you want the product to do. Remember, product creation is a step-by-step process, each one as important as the last. If you don't have a vision regarding the product you intend to make, you are more likely to get sidetracked and not achieve what you set out to do. There are some questions to keep in mind regarding the vision you have for your product:

- What are your goals?

- What do you want to achieve?

- How will you know if you have achieved your goals?

- Is your idea in line with what the customer wants/needs?

After you have answered these questions, you should have a clearer vision of your product and what you want it to achieve.

Identify the Team Members:

The next step is to identify who will be working on the product. This includes the development team, the marketing team, and any other stakeholders

Define the target market: The second step is defining the product's target market. This step is important because it will help determine the features that need to be included in the product and how it should be marketed. There are a few questions that should be answered when defining the target market:-

Who is the product for?

- What needs does the product address?

- What are the demographics of the target market?

- What is the psychographics of the target market?

Establish Your KPIs:

Key Performance Indicators (KPIs) are a way to measure the success of your product. In order to establish KPIs, you need first to identify what success looks like for your product. Once you have done that, you can establish KPIs that will help you track and measure that success. Some examples of KPIs are:-

Number of active users

- Customer satisfaction scores

- Net Promoter Score (NPS)

- Churn rate- Revenue- Conversion rate

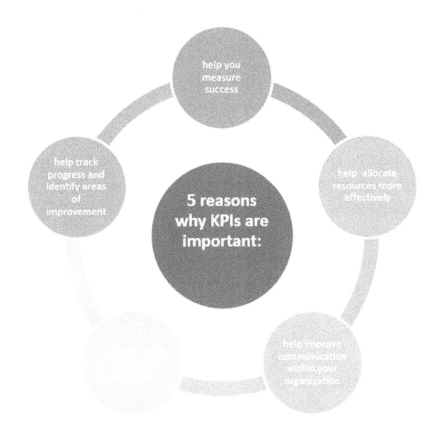

Understand the User's Needs

Once the target market has been defined, the next step is to understand the user needs. This can be done by conducting market research or by talking to potential customers. The goal here is to understand what the users want and need from the product. Some questions that should be answered during this step are:-

- What are the user needs?

- What are the user pains?

- What are the user gains?

- What are the user jobs?

- What are the user journeys?

Lastly, you need a rough product plan:

After you understand the ***what and why*** of the product, it's time to start thinking about the how. In other words, you need to create a rough plan for how the product will be developed. This includes deciding on the features that will be included in the product, the schedule for development, and the budget. Creating a great product doesn't have to be complicated. By following these simple steps, you can ensure that your product is well-defined and meets the needs of your users. By following these simple steps, you will be on your way to designing great products!

Customer Research

After the kick-off meeting, the next step is to conduct customer research. This involves talking to potential customers and understanding their needs. This step is important because it will help determine the features that need to be included in the product and how it should be marketed.

Creating User Stories

After the customer research has been conducted, the next step is to create user stories. User stories are short descriptions of how the product will be used. They help define the product's functionality and how it will meet the users' needs.

Here are a few user stories:

- *As a user, I want to be able to order food from restaurants so that I can have it delivered to my doorstep.*

- *As a user, I want to be able to find restaurants near me so that I can order food from them.*

- *As a user, I want to be able to see the menu of a restaurant so that I can order food from them.*

- *As a user, I want to be able to pay for my food so that I can receive it.*

- *As a user, I want to be able to track my order so that I can know when it will arrive.*

These user stories define the product's functionality and how it will meet the users' needs.

Creating a Prototype

The next step is to create a prototype of the product. This will help to visualize the product and how it will work. It is important to note that the prototype should not be too detailed. The goal is just to create a basic version of the product. For example, the prototype might be a wireframe if you are creating a website. Concept mapping is also a useful tool to use during the prototype phase.

According to the Nielsen Norman Group, concept mapping is "a visual way to organize and understand relationships between ideas."

What are Some Methods to Create Prototypes?

Creating a Low-Fidelity Prototype:

A low-fidelity prototype is a quick and easy way to create a prototype. This type of prototype is usually created using paper and pencil. Low-fidelity prototypes are often used to test basic concepts. According to the former Google design ethicist Tristan Harris, low-fidelity prototypes are "good for exploring a space of possible solutions." Low-fi prototypes are best for getting initial feedback on structure and layout before investing too much time in details. At this point, you don't really need to create high-fidelity prototypes. You just want to get a feel for the product and how it will work.

Your goal is to create a prototype that is:

- Quick and easy to make

- Focuses on the essential features of the product

- Allows you to test basic concepts

Tools such as storyboards, post-it notes, and mind maps can be used to create low-fidelity prototypes. All of these techniques are quick and easy to learn. According to the Interaction Design Foundation, "the main advantage of using low-fidelity techniques is that they require very little time and effort to produce."

Creating a High-Fidelity Prototype:

A high-fidelity prototype is a more detailed version of the product. This type of prototype is usually created using software such as

Adobe Photoshop or Sketch. High-fidelity prototypes are often used to test more complex concepts.

Creating an Interactive Prototype:

An interactive prototype is a prototype that allows users to interact with it. This type of prototype is usually created using software such as Adobe Flash or HTML. Interactive prototypes are often used to test the usability of the product.

The goal is to visually represent how the product will look and work. This will help to communicate the idea to others and get feedback on the design.

Pro tip: You can also use online tools such as Balsamiq to create low-fidelity prototypes.

Chapter 5

Product Research and Designing

In this chapter, we will discuss:

- What is product research?

- How to carry out product research

- What are the methods through which we can carry out product research?

- What is product designing?

- Methods through which we carry out effective product designing

We need to stop worrying about proving the value of design
& just focus on outcomes that provide value. – Denis Weil

Once you understand the product and the user, it's time to move on to the product research phase. Many books have this stage intertwined with the first one since most product developers believe product research is carried with the product definition. However, we've decided to keep it separate to avoid confusing the two.

Product research is the process of gathering data about a product, its users, and the market. This data is then used to make informed decisions about the product. Together, these stages help answer the question: ***what should be built?*** It is important to note that product research is done via market and user research.

Conducting Market Research

Market research is the process of gathering data about the market in which the product will be sold. This data can be used to understand the needs of the market and the competition. There are many different ways to conduct market research, but some of the most common methods include surveys, interviews, and focus groups.

Conducting User Research

User research is the process of gathering data about the users of the product. This data can be used to understand the needs of the users and how they interact with the product. There are many different ways to conduct user research, but some of the most common methods include surveys, interviews, and focus groups.

Why Is Research Important?

Product research is important because it helps to inform the product development process. Without research, it would be difficult to understand what the product should be, who the users are, and what the market needs. Research is essential to developing a successful product.

The product research phase is important because it helps to answer the question: what should be built? This phase is important because it helps to understand the needs of the users and the market. According to Steve Blank, "The best way to learn about your customers is to talk to them." Product research is essential to developing a successful product.

The famous Steve Jobs quote, *"Design is not just what it looks like and feels like. Design is how it works."* perfectly sums up the importance of research in product development. A great product is not only aesthetically pleasing but also easy to use and solves a problem for the user. All of this is only possible if the product has been researched thoroughly.

Don't worry if you have not done your market research or user research. For that matter, for your product, tools such as market segmentation, target market, and persona development can help you out.

Product definition and *product research* are two important stages in the product development process. Product definition helps to answer the question: what is to be built? And product research helps to answer the question: what should be built? Together, both of these stages help to develop a successful product.

We personally think the game is already won if you know your market and users inside out and what their needs are. But if you want to increase your chances further, read on to find our top 3 tips for product definition and product research.

According to us, these 3 tips are absolutely essential for any product development team.

1. Define the problem that you are solving:

The first and most important step is to define the problem that you are solving. This may seem like a no-brainer, but many product development teams forget to do this. Without a clear problem definition, moving forward with the development process is difficult.

2. Identify your target market:

The next step is to identify your target market. This is important because it will help you understand your target market's needs. Once you know your target market, you can start to develop a product that meets their needs.

3. Conduct market and user research:

Once you have defined the problem and identified your target market, the next step is to conduct market and user research. This research will help you to understand the needs of your users and the market. Without this research, it would be difficult to develop a product that meets the needs of your users

What features or entities must you keep in mind while building your market segmentation report?

Some of the features or entities that you need to keep in mind while building your market segmentation report are:

- Age

- Gender

- Location

- Income

- Occupation

- Marital status

- Interests

- Buying habits

McDonald's is one of the world's most well-known fast food chains. They have successfully used market segmentation to grow their business. For example, it segments its market by age, gender, location, income, and occupation. This has helped them to develop products and marketing programs that meet the needs of their target market. But how do they segment their happy meals for kids?

By age and gender. They have different happy meal options for boys and girls. This helps to ensure that kids are getting the products that they want. McDonald's also offers discounts and coupons to their target market. This helps to attract more customers and grow their business.

Survey

A survey is a questionnaire that you can use to gather data from your target market. Surveys are a great way to collect data about your target market's needs, wants, and behaviors. According to HubSpot, "*A survey is a research method used to collect data from a group of people. Surveys are usually administered either in person, by phone, or online*."

What are some of the benefits of conducting a survey?

Some of the benefits of conducting a survey are:

- Helps you to understand your target market

- Helps you to collect data about your target market's needs, wants, and behaviors

- Helps you to develop products that meet the needs of your target market

- Helps you to create marketing programs that appeal to your target market

- Helps you to understand your target market's buying habits

- Helps you to segment your target market

When you are conducting surveys, here are a few tips and points to keep in mind:

- Keep your questions short and to the point

- Ask closed-ended questions that can be answered with "yes" or "no."

- Avoid asking leading questions

- Make sure your questions are clear and easy to understand

- Test your survey on a small group of people before you launch it

In 2016, Nike wanted to introduce a new line of shoes. They conducted a survey to gather market data from their customers. The survey asked customers about their current Nike shoes, their favorite style of Nike shoes, and what they would like to see in a new line of Nike shoes. The survey helped Nike understand its target market's needs and develop a new product that would meet those needs.

The famous market researcher Philip Kotler once said, "The aim of marketing is to make selling superfluous." In other words, marketing should create a need or want in the customer's mind, not selling a product.

Think about it this way: if you're selling a new type of toothbrush, your goal shouldn't be to sell as many toothbrushes as possible. Your goal should be to create a need or want for that toothbrush in the minds of your target market. Once you've done that, the selling will take care of itself.

Creating a need or want in the customer's mind is called *creating demand*. And there are a few ways that you can create demand for your product or service:

- You can create demand by offering a new product or service that meets a need or want that your target market has.

- You can create demand by improving upon an existing product or service.

- You can create demand by making your product or service more accessible to your target market.

- You can create demand by making your product or service more affordable to your target market.

Conducting Heuristic Evaluations

A heuristic evaluation is a method of usability testing where a small group of people test a product or service. Heuristic evaluations are often used to test websites or apps.

To conduct a heuristic evaluation, you will need to gather a group of people who are familiar with the product or service you're testing. You will then ask them to use the product or service and look for any problems. These problems can be anything from user experience issues to technical problems.

Once you have collected all of the data, you will need to analyze it to understand the problems. This will help you to improve the product or service. Heuristic evaluation is a good benchmark against which other usability testing methods can be compared. According to Jakob Nielsen, heuristic evaluation is "the most cost-effective way to conduct usability testing for interactive systems."

Jakob Nielsen's 10 Heuristics for User Interface Design:

1. *System's status Visibility:* The service or product should always keep customers up-to-date about what is going on through graphic indications, audio signs, or manuscripts.

2. *Match between system and the actual world*: The service or product should express the users' linguistics, use familiar metaphors, and make information easy to understand.

3. *User control and autonomy*: Users should be able to undo actions if they make a mistake.

4. *Uniformity and principles*: The system should use consistent language and icons and follow standard conventions.

5. *Error avoidance*: The system should prevent errors or help users recover from them.

6. *Recognition rather than recollection*: The product should diminish the user's recollection burden by making information visible and easily accessible.

7. ***Flexibility and productivity of use***: The system should be flexible so that users can tailor it to their needs. It should also be efficient so that users can accomplish their goals quickly and easily.

8. ***Artistic and simple design***: The system should be aesthetically pleasing and use only as much information as is necessary.

9. ***Assist users in identifying, detecting, and recovering from errors***: The product or service should assist users in identifying, analyzing, and recuperating from mistakes.

10. ***Support and documentation***: The system should come with help and documentation so that users can learn to use it.

When it comes to product usability, there are a few real checks and tips you need to keep in mind:

- Make sure you have a great team who are all on the same page when it comes to the product development process.

- Use heuristic evaluations to test products and services.

- Make sure you have a clear understanding of your target market.

- Use surveys and customer data to understand the needs of your customers.

- Use consistent language and icons throughout the product.

- Prevent errors and help users recover from them.

- Make sure the product is aesthetically pleasing.

- Provide help and documentation so users can learn to use the product.

UX Design isn't easy, but following these tips will help you create great products that your customers will love.

The Apple iTunes Heuristic Evaluation Report is a testing method used to identify issues with the user interface of the iTunes software. The report is based on Jakob Nielsen's 10 Heuristics for User Interface Design.

The report found that iTunes was generally easy to use and had a consistent interface. However, there were some areas where improvements could be made. For example, the report found that iTunes could improve its visibility of system status, use of consistent language and icons, and error prevention.

Overall, the Apple iTunes Heuristic Evaluation Report is a useful tool for identifying potential problems with the user interface of iTunes. By following the recommendations in the report, iTunes can improve its usability and make it even easier for users to manage their music collections.

Interestingly you can also categorize your usability issues as low, medium, or high severity.

- **Low severity** means that the issue is not likely to cause any problems for users.

- **Medium severity** means that the issue could cause some confusion or inconvenience for users.

- **High severity** means that the issue is likely to cause serious problems for users.

For example, a low severity issue might be that the text on a button is slightly unclear. A medium severity issue might be that users have to scroll through a long list of options to find the one they want. A high severity issue might be that the wrong information is displayed to users or that users are unable to complete a task because of a problem with the interface.

User Research Report

User research is a process of investigating how people use products or services in order to understand their needs and wants. User research can inform the design of new products or services or improve the usability of existing ones.

There are many different user research methods, but all share the same goal: to gain insights into how people think, feel, and behave. Some common methods of user research are interviews, focus groups, surveys, and usability testing. According to the Nielsen Norman Group, interviews are the most popular method of user research, followed by surveys, focus groups, and usability testing.

According to Jon Kolko, a famous design thinker, user research is the most important part of the design process.

No matter what method you use, user research should be an iterative process conducted throughout the entire design process. It should not be limited to the beginning or the end but should be ongoing in order to gather feedback and improve the design continuously.

User research can be conducted by anyone with an interest in understanding how people use products or services. However, it is often conducted by professional user researchers with specific training and skills.

In 2015, the online retailer Amazon launched a new line of products called AmazonBasics. The AmazonBasics were a line of everyday items, such as batteries and HDMI cables, that were sold at a lower price than similar products from other brands.

To ensure that AmazonBasics was designed to meet the needs of its customers, Amazon conducted extensive user research. This included surveys, interviews, focus groups with customers, and usability testing of the products.

The user research was successful in identifying a number of areas where AmazonBasics could be improved. For example, the products were not always clearly labeled, and the packaging was sometimes confusing. As a result of the user research, Amazon made a number of changes to AmazonBasics, such as improving the labeling and packaging.

The user research was successful in helping Amazon to create a successful line of products. AmazonBasics are now one of the best-selling lines of products on Amazon and have helped to establish Amazon as a major player in the retail market. User research is an essential part of the design process and can help to create successful products. In the case of AmazonBasics, user research was key to the success of the product line.

User research can be conducted at any stage of the product development process, from early concept development to post-launch evaluation.

Why Is User Research Important?

User research is important because it helps ensure that products or services are designed to meet the needs and wants of the people who will use them.

User research can help to:

- Identify user needs and requirements
- Improve the usability of products or services
- Reduce development costs by avoiding mistakes
- Increase customer satisfaction and loyalty

What Are Some Common Problems With User Research?

Common problems with user research include:

- **Inadequate planning**: User research should be planned in advance in order to ensure that the right questions are asked, and the correct data is collected.

- *Insufficient resources*: User research requires time, money, and personnel. If these resources are not available, the quality of the research will suffer.

- *Lack of expertise*: User research should be conducted by trained professionals who know how to collect and interpret data properly.

- *Bias*: User research can be biased by the researcher's own personal preferences or preconceptions.

- *Social desirability bias*: People may give responses that they think are socially desirable rather than honest ones.

- *Hawthorne effect:* People may change their behavior simply because they know they are being observed.

- *Experimenter's bias*: Researchers may inadvertently influence the results of the research by the way they collect or interpret data.

How can user research be conducted effectively?

User research can be conducted effectively by following these steps:

1. **Plan the research**: Decide what you want to learn and how you will go about collecting data.

2. **Collect data**: Use a variety of methods to collect data from users, such as interviews, focus groups, surveys, and usability testing.

3. **Analyze data**: Carefully analyze your collected data, looking for patterns and insights.

4. **Share findings**: Present your findings to stakeholders in a clear and actionable way.

User research is important for understanding how people use products or services. When conducted effectively, it can help to improve the usability of products or services and increase customer satisfaction.

Here we say *goodbye* to the product research phase and move on to the next stage, which is designing the product. Till now, we have everything we need regarding the product. We have our findings and market data regarding the product, so now we will start making prototypes of our product!

Product Design

In this stage, the team creates a prototype of the product. And might we say these prototypes are actually high-level working prototypes of the product or what the product will look like? This is usually done through a software program like Sketch or Photoshop. According to the famous industrial designer Raymond Loewy "You can sell anything if you package it right." This is what packaging means in the product design stage.

The second stage is all about the ***user experience,*** or what we like to call ***UX***. This is where you start to think about how the user will interact with your product. You want to make sure the user experience is simple and easy to use. You don't want to make it too complicated, or the user will get frustrated and give up.

Creating a prototype helps the team understand how the product works and what users will see when interacting with it. There are many, many ways to create prototypes. You may begin with low-fidelity prototypes, which are basically sketches on a whiteboard or paper. Or you may go straight to high-fidelity prototypes, which look and feel almost like the real thing.

So how does prototyping works? Well, you may start off with a simple sketch. Let us see how UX designers make visual designs of products:

Sketching

This is the first and most important step in the UX design process. A sketch is a quick drawing representing your product's user interface. It can be done by hand or using software like Photoshop or Sketch. When you are making a sketch of your product, you should think about the following:

- What are the most important elements of the user interface?

- How will users interact with the product?

- What is the overall look and feel of the product?

According to **Smashing Magazine**, "A sketch is worth a thousand words." This means that a sketch can communicate a lot of information about the product in a very short amount of time. A good sketch will help the team understand the product and make better decisions about the design.

Here are some tools and techniques that will help you create awesome sketches:

- *Brainstorming*: This technique can be used to generate ideas for the product. It can be done individually or in a group.

- *Mind mapping*: This technique can be used to organize ideas. It involves creating a map of the ideas, with the main idea in the center and related ideas branching off from it.

- *Storyboarding*: This technique can be used to create a visual representation of the user journey. It involves creating a series of images showing the user's steps when using the product.

- *Wireframing*: This technique can be used to create a skeletal representation of the user interface. It involves creating a simple version of the product that includes the most important elements of the user interface.

After you have made a sketch of your product, you can then begin to create a prototype. A prototype is a working model of your product. It can be used to test your product's usability and get user feedback. There are many different types of prototypes, ranging from low-fidelity to high-fidelity.

Low-fidelity prototypes: These prototypes are usually quick and easy to create. They are often made with paper or simple software programs. Low-fidelity prototypes are good for testing the overall concept of the product.

High-fidelity prototypes: These prototypes are more realistic and closer to the final product. They are often made with high-end software programs or 3D printers. High-fidelity prototypes are good for testing the details of the product.

Wireframing:

This technique can be used to create a skeletal representation of the user interface. It involves creating a simple version of the product that includes the most important elements of the user interface. Digital wireframing tools like Balsamiq and Wireframe.cc make it easy to create wireframes. Apart from these tools, you can also use pen and paper to create wireframes. According to the famous UX firm, Hteo, *"Wireframing is the backbone of the UX design process."* The Consultation firm HeadScape, says that wireframing is an "essential step" in their design process.

The benefits of creating wireframes:

- Helps you to think about the structure of the user interface
- Helps you to communicate your ideas to others
- Helps you to make changes to the design quickly and easily

- Helps you to focus on the most important elements of the user interface

Once you have a low fidelity or even a high-fidelity wireframe, you can move on to creating the visual design of your product. Visual design is the process of creating the look and feel of your product. It involves choosing the colors, fonts, and images that will be used in the product.

Some tips for creating great visuals:
- Use colors that are appropriate for the brand and the product

- Use fonts that are easy to read

- Use images that are relevant to the product

Apple is known for its simple and elegant designs. The company's products are known for their minimalistic aesthetic. This aesthetic is also reflected in Apple's wireframes.

High-fidelity wireframes have their own advantages, but low-fidelity wireframes have some advantages over them. Low-fidelity wireframes are quicker and easier to create. They are also more flexible and easier to change. Low-fidelity wireframes are good for early-stage design when the product is still in flux and subject to change.

Some tips for creating great wireframes:
- Keep it simple

- Focus on the essential elements

- Make it easy to change

- Use paper or digital tools

Creating a great product is not easy. It requires careful planning and design. However, by following the tips above, you can create a simple and effective product.

Mockups:

A mockup is a static representation of the product. It can be used to test the visual design of the product. Mockups are usually created in Photoshop or other design programs.

Some tips for creating great mockups:

- Use real photos or illustrations

- Use the actual size of the product

- Place the mockup in a realistic environment

- Use shadows and lighting to make the mockup look realistic

Many companies, like Google and Facebook, use mockups to test the visual design of their products. According to the famous UX designer Luke Wroblewski, "mockups are a key part of the design process." Your mockups have the power to make or break your product, so it's important to create them carefully. Tools such as Balsamiq and Photoshop make it easy to create mockups.

This is all great, but do companies make mockups, wireframes, and sketches, or do they choose any or a mixture of these?

It's important to understand that there is no one right way to design a product. Different companies use different methods, depending on their needs. For example, some companies might create wireframes first, followed by mockups. Others might start with sketches and then move on to wireframes and mockups. There is no wrong way to design a product, as long as the end result is a great product.

According to The CEO of *InVision*, **Casey Winters**, *"Design is a process, not a single step."* He says that the important thing is to "iterate and get feedback early and often."

The most important thing is to create a great product. The method you use to design the product is less important than the end result.

A question arises here, should designers spend hours creating high-fidelity mockups? The answer is no. It is more important to focus on the user experience and the functionality of the product than on the visuals. The visual design can always be improved later.

The goal of design is to create a product that is both useful and usable. A product that is useful but not usable is not a good product. A product that is usable but not useful is also not a good product. The best products are both useful and usable.

Some tips to remember when creating mockups:

- Use real photos or illustrations
- Use the actual size of the product
- Place the mockup in a realistic environment
- Use shadows and lighting to make the mockup look realistic

- Focus on the user experience and functionality, not on the visuals. The visual design can always be improved later.

Prototyping:

Prototyping is the process of creating a working model of the product. It is used to test the design's feasibility and determine if the product is user-friendly. Prototypes can be created using paper, digital tools, or even 3D printers. So what is the difference between wireframes, mockups, and prototypes?

- Wireframes are low-fidelity, static representations of the product. They focus on the content and the structure of the product.

- Mockups are high-fidelity, static representations of the product. They focus on the visual design of the product.

- Prototypes are working models of the product. They focus on the user experience and functionality of the product.

"The goal of prototyping is to test the design of the product and to find out if it is user-friendly. Prototyping is a key part of the design process."

What are some famous prototyping tools?

Some of the most popular prototyping tools are InVision, Justinmind, and Adobe XD.

What are some tips for creating great prototypes?

- Use real photos or illustrations

- Use the actual size of the product

- Place the mockup in a realistic environment

The company Google is well-known for its use of prototypes. The company creates a prototype for every new product or feature it develops. Prototypes are used to test the design's feasibility and determine if the product is user-friendly. Google has even been known to release prototypes to the public to get feedback before the product is released.

When should you create a prototype?

The best time to create a prototype is when the design is complete. Prototypes should be created before the product is developed.

What are some benefits of prototyping?

These are some benefits of prototyping:

- Prototyping can help to reduce the risk of failure.

- Prototyping can help to improve the quality of the product.

- Prototyping can help to save time and money.

- Prototyping can help to improve communication between team members.

- Prototyping can help to reduce the chances of scope creep.

Should a company go far with low-fidelity prototypes or high-fidelity?

There is no right or wrong answer to this question. It depends on the needs of the project. If the goal is to test the feasibility of the design, then a low-fidelity prototype may be sufficient. If the goal is to test the user experience and functionality of the product, then a high-fidelity prototype may be necessary.

Marty Cagan, the famous product executive, has a great saying about prototypes: *"If you haven't created a prototype, you haven't really designed anything."*

Creating prototypes is a key part of the design process. Prototypes help to reduce the risk of failure, improve the quality of the product, and save time and money.

According to Smashing Magazine, "A good prototype is one that allows you to test the design and find out if it is user-friendly. Prototypes should be created before the product is developed."

Here are some tips for creating prototypes:

- Use real photos or illustrations

- Use the actual size of the product

- Place the mockup in a realistic environment

- Use shadows and lighting to make the mockup look realistic

- Focus on the user experience and functionality, not on the visuals. The visual design can always be improved later.

High-fidelity and low-fidelity prototypes both offer advantages and disadvantages. It's important to choose the right fidelity for the task at hand.

What Other Design Specifications Documents Does the Development Team Need to Carry Out During Product Design?

- The development team will need a functional specification document that outlines the product's functionality.

- The development team will need a technical specification document as well as user flow diagrams, which map out the user's journey through the product.

A user flow diagram is a map of the user's journey through the product. User flow diagrams can be used to help design the user experience and improve the product's usability. To provide ease of use, Developers and designers must take into account user flow diagrams. In fact, a user flows diagram is a great tool to achieve UX efficiency.

- The development team will also need a design specification document outlining the product's visual design.

Style Document

A style document is a document that defines the visual style of the product. The style document should include the product's color palette, typography, and iconography.

What is the goal of a style document?

The goal of a style document is to define the product's visual style. The style document should include the product's color palette, typography, and iconography.

Chapter 6

Product Development Phase

In this chapter, we will discuss:

- What is a product's development phase?

- What activities are carried out in this stage?

- What are agile development methods?

- What are sprints and backlogs?

- What are the main activities that take place during the development phase of a product?

- 6 stages of product development

Culture, leadership & employee engagement are the essentials for a great customer experience. – Steve Cannon

Now that we have covered a product's planning, research, and designing stages, let us move on to the product development phase. The development phase is where a product is actually built and made ready for release. This is the stage where all the coding, programming, and testing are done. A product development team generally consists of software engineers, programmers, testers, etc. After months or even years of planning and researching, it all comes down to this one moment- release. All your efforts, all your team's efforts, and all the company's resources are put to use to make the product successful. All the stakeholders are involved in this stage, and the pressure is high. A lot of documents are created or updated in this stage, for example- product requirement documents, design documents, test cases, user stories, etc.

Your previous designs, prototypes, requirements, and documentation created in the earlier stages come in handy during development. They act as a guide for the developers and help them understand your vision for the product. The developers use various programming languages and tools to build the product according to the designs. The development team also takes help from previous feedback to make sure that they are on the right track.

And since in today's world, all the development is done through the agile model, the development phase is divided into sprints. In each sprint, the developers work on a few features and deliver them at the end of the sprint. Each sprint is followed by a review meeting in which stakeholders give their feedback about the progress made and what can be improved.

The *agile model* has helped *product development* become more *efficient* and *effective*. It has made it easier to track the progress and also to make changes according to the feedback. The development phase can be a very stressful time as there is a lot of pressure to deliver a successful product. But with proper planning, research, and designing, it can be made easier.

Once the product is developed, it needs to be tested thoroughly in order to check if it is working as intended and if there are any bugs or errors.

What are some of the documents created in this stage?

- Product requirement document

- Design document

- Test cases

- User stories

Phases of the Development Stage

The development phase or implementation phase isn't just a standalone phase. As we discussed, you must work in iterations, in circles. This is why you need to keep going back to the previous stages. For this reason and for ease of simplicity and effectiveness, the development team works in two phases in the development phase:

1. **Coding and programming**

2. **Testing**

> o Coding and programming is the process of writing the code for the product. The developers use various programming languages to write the code. They also

take help from various tools and libraries to ensure that the code is high quality.

o Testing is the process of testing the product for bugs and errors. The testing team uses various techniques to test the product. They also use various tools to automate the testing process.

Both of these parts are essential in order to create a successful product. In the *programming part*, the actual product is built and made ready for release. This is where all the coding, programming, and testing are done. And in *the testing part*, the product is put through different tests to check if it is working as intended and if there are any bugs or errors. Beta testing is done at this stage to get feedback from real users before the product is finally released. Although testing and launch are sometimes made into separate stages, it is generally considered to be part of the development phase.

However, we will discuss the testing phase as a separate stage from this one. For now, we will concentrate on the development stage.

What are some testing techniques carried out in the development stage?

- **Unit testing**: This is a type of testing where individual units or components of the software are tested.

- **Integration testing**: This is a type of testing where the various units of the software are integrated and tested as a whole.

- **System testing**: This is a type of testing where the entire system is tested.

- **Acceptance testing**: This is a type of testing where the client or the end user tests the product to see if it meets their requirements.

Now the question arises here what exactly does the development team do in this stage? How do they carry out their tasks? The development team builds the product according to the designs and specifications created in the previous stages. They use various programming languages and tools to build the product. The development team also takes help from the feedback gathered in the earlier stages to make sure that they are on the right track. As we discussed before, sprints are generally used in product development.

This helps the developers to track their progress and make necessary changes. Thanks to sprints, the development process becomes more efficient and effective.

That's all good. But why sprints? Why are they important for a product, especially a UX-driven product?

Well, sprints do have their own sets of benefits. Sprints help in the following ways:

- They help in dividing the work into manageable chunks so that the developers can focus on one thing at a time.

- They help in tracking the progress and also make it easier to make changes according to the feedback.

- They help in making the development process more efficient and effective.

Sprints are generally used in product development because they:

- Make it easier to make changes based on feedback

A sprint is a period of time (usually 2-4 weeks) in which a specific set of tasks is completed. At the end of each sprint, the product is released to the end users.

The development stage also works in *iteration*. The first task of any development team is to **sit with the client** and understand their requirement. After that, they start working on a product design document. This **document** contains all the specifications of the product, like the *features*, *user flow*, etc. Once the design is approved by the client, the **development** team starts working on the actual product. They use various **programming languages** and **tools** to build the product. The **development** team also takes help from the feedback gathered in the earlier stages to make sure that they are on the right track. As we discussed before, *sprints are generally used in product development*. In each sprint, the developers work on a few features and deliver them at the end of the sprint. After each sprint, there is a review meeting in which stakeholders give their feedback about the progress made and what can be improved. But since this is iteration work, you may have to go back to the board and take the client's feedback on every milestone.

A lot of activities and a lot of documentation are carried out in this phase, Let us discuss the activities one by one, and then we'll move on to the documentation phase:

Four main activities take place during the development phase of a product:

1. **Designing:** In this stage, the product is designed according to the specifications and requirements. This includes creating the user interface, wireframes, and prototypes. For instance, if you are designing a website, then this stage would involve creating mockups and prototypes of the website.

2. **Coding**: In this stage, the actual coding of the product takes place. The developers use various programming languages to write the code for the product. They also take help from various tools and libraries to ensure that the code is high quality.

3. **Testing**: In this stage, the product is tested for bugs and errors. The testing team uses various techniques to test the product. They also use various tools to automate the testing process.

4. **Documentation**: In this stage, the product is documented. This includes creating the user manual, installation guide, and other technical documents. The documentation helps the users understand the product better and helps the developers track the changes made to the product.

Here I am going to drop a word called *"backlogs." **Backlogs*** are mostly created in the *development phase*. Before delving deeper into these concepts, I would like to clarify what they *actually* are **backlogs**.

Backlogs are lists of tasks *that need to be completed*. They help teams track their progress and ensure they are on track to meet their goals. Backlogs can be used for any type of project, from product development to software engineering. For instance, a product development team may have a backlog of features that they want to add to their product. A software engineering team may have a backlog of tasks they need to complete to finish a project. Backlogs help teams to stay organized and efficient.

How do backlogs assist various activities in the development phase?

The first task in the development phase is to create a **product backlog**. A product backlog is a prioritized list of all the features that need to be added to a product. This list is created based on input from various stakeholders, such as the product owner, the development team, and the end users.

The next task is to create a ***sprint backlog***. A sprint backlog is a list of all the tasks that need to be completed in a sprint. This list is created based on the product backlog.

The development team then starts working on the tasks in the ***sprint backlog***. They work on one task at a time and move on to the next task only when the first task is completed.

Once all the tasks in the sprint backlog are completed, the product is then ready for release.

There are many different methods of software development, but the most common one is agile development. Agile development is a process that allows for small, incremental changes to be made to a product over time. This makes it easier to track progress and make changes as needed.

To help you understand how backlogs work, let's take the example of amazon.com

The product development team at amazon.com has a backlog of features that they want to add to their product. In order to prioritize these features, the team uses a user story mapping technique. This helps them understand their users' needs and prioritize features that will address those needs. The team then creates a roadmap that outlines the order in which they will add these features to their product. This helps them stay on track and ensures that they progress towards their goals.

Up till now, you have been reading all the good about the development stage. Heck, you are probably wondering how easy it is to create a product. But let me tell you, it's not all rainbows and butterflies. The development stage is full of challenges and obstacles. And the biggest challenge is to handle the client's expectations. Because at the end of the day, it is the client who is going to use the product. So their feedback is really important. And sometimes, you may have to go back to the drawing board and make changes according to their feedback. But that's okay because that's what the agile model is all about—making changes according to the feedback.

What are some of the challenges faced in this stage?

- There is a lot of pressure to deliver a successful product.

- The development process can be very lengthy and time-consuming.

- It can be difficult to track the progress and make changes according to the

What are some of the documents created in this stage?

- Product requirement document

- Design document

- Test cases

- User stories

The 6 Stages of Product Development

Okay, now that we have some basic knowledge of the development stage, let us take things to a *whole new level*. Like the other stages of product creation, the development phase also goes through some stages or steps. You may also call these stages as *"product life cycle."*

1. Ideation

2. Research

3. Planning

4. Designing

5. Coding

6. Testing

7. Documentation

8. Release

9. Maintenance

Each stage of the development process is important and has its own set of tasks that need to be completed. Skipping any one of these stages can lead to a sub-par product that will not be able to meet the needs of the end users.

It is important to note that the development phase is an iterative process. This means that the product will go through each of these stages multiple times before it is finally released to the end users.

Ideation is the first stage of product development. This is when the product is first conceived. The team comes up with an idea for a new product or feature. They then do some initial research to see if the idea is feasible. This stage is all about generating new ideas and seeing if they are worth pursuing.

What are some activities that help carry out ideation for the team?

- *Brainstorming*: This is a technique where the team comes up with as many ideas as possible. They can either do this individually or as a group.

- *Mind mapping*: This is a technique where the team maps out their ideas visually. This helps them see the connections between different ideas and develop new ones.

- *SWOT analysis*: This is a technique where the team looks at the strengths, weaknesses, opportunities, and threats of an idea. This helps them to assess the risks and benefits of pursuing it.

Research is the second stage of product development. This is when the team deep dives into the problem they are trying to solve. They gather data and feedback from different sources. This helps them validate their ideas and develop a solution that is most likely to be successful.

What are some activities that help carry out research for the team?

- *Surveys*: This is a technique where the team sends out surveys to their target audience. This helps them gather data about the problem they are trying to solve.

- *Interviews*: This is a technique where the team interviews their target audience. This helps them to understand the problem from their perspective and gather feedback about potential solutions.

- *Competitor analysis*: This is a technique where the team looks at what their competitors are doing. This helps them understand the market and come up with a solution different from what is already out there.

Planning is the third stage of product development. This is when the team creates a plan for how they are going to achieve their goals. They create a roadmap outlining the different features and milestones they want to achieve. This helps them stay on track and make sure they are making progress.

What are some activities that help carry out planning for the team?

- *Road mapping*: This is a technique where the team creates a roadmap that outlines the different features and milestones that they want to achieve. This helps them stay on track and make sure they are making progress.

- **Task management**: This is a technique where the team uses a tool like *Jira* to track their progress. This helps them to see what tasks need to be done and who is working on them.

- **Scrum**: This is a technique where the team uses a tool like *Trello* to track their progress. This helps them to see what tasks need to be done and who is working on them.

Designing is the fourth stage of product development. This is when the team creates the user interface and user experience for their product. They use different design principles to create a product that is both visually appealing and easy to use.

What are some activities that help carry out designing for the team?

- **User research**: This is a technique where the team does research on their target audience. This helps them understand their needs and develop a design that meets their expectations.

- **Mockups**: This is a technique where the team creates a mockup of their product. This helps them to get feedback about the design and make changes before they start development.

- **Prototyping**: This is a technique where the team creates a prototype of their product. This helps them test the design and ensure it is user-friendly.

Development is the fifth stage of product development. This is when the team starts to build the product. They write code and create the different features that they have designed. This is an iterative process where the team makes changes and improves the product based on feedback from users.

What are some activities that help carry out development for the team?

- *Version control*: This is a technique where the team uses a tool like Git to track their progress. This helps them to see what changes have been made and who made them.

- *Continuous integration*: This is a technique where the team uses a tool like Jenkins to build and test their code automatically. This helps them to find errors quickly and fix them before they cause problems.

- *Documentation*: This is a technique where the team writes documentation for their product. This helps users understand how the product works and helps developers know what needs to be done.

Testing is the sixth stage of product development. This is when the team tests their product to make sure that it is working as intended. They use different techniques like unit testing and user acceptance testing to find bugs and fix them.

What are some activities that help carry out testing for the team?

- *Unit testing*: This is a technique where the team writes code to test the functionality of their product. This helps them to find bugs and fix them before they cause problems.

- *User acceptance testing*: This is a technique where the team asks users to test their product. This helps them to find bugs and fix them before they cause problems.

- *Performance testing*: This is a technique where the team tests their product to see how it performs. This helps them find issues that need to be fixed to improve performance.

- *Integration testing*: This is a type of testing where the various units of the software are integrated and tested as a whole.

- *System testing*: This is a type of testing where the entire system is tested.

Deployment is the seventh stage of product development. This is when the team puts their product into production and makes it available to users. They use different techniques to deploy their product, like blue/green deployment and canary release.

What are some activities that help carry out deployment for the team?

- *Blue/green deployment*: This is a technique where the team deploys their product to two different environments. This helps them test the product's new version in production before making it available to all users.

- *Canary release*: This is a technique where the team first releases a new version of their product to a small group of users. This helps them test the new version of the product and ensure that it is stable before making it available to all users.

- *A/B testing*: This is a technique where the team releases two different versions of their product to two different groups of users. This helps them compare the two versions and see which is better.

The eighth and final stage of product development is ***maintenance***. This is when the team provides support for their product and makes sure that it is up to date. They use different techniques to maintain their product, like patching and versioning. What are some activities that help carry out maintenance for the team?

- *Patching*: This is a technique where the team fixes bugs in their product. This helps them keep their product up to date and ensure that it is stable.

- *Versioning*: This is a technique where the team releases new versions of their product. This helps them to add new features and improve existing ones.

- *Monitoring*: This is a technique where the team monitors their product. This helps them see how it is used and identify issues that need to be fixed.

The agile development process is flexible and can be easily adapted to changing needs. It is also very efficient and helps ensure that products are high quality.

If you want to learn more about agile development, many resources are available online. You can also attend training courses or workshops offered by organizations such as the Agile Alliance.

Product Testing and Launch Stage

After all the hard work in designing, developing, and testing the product comes the moment of finally launching it in the market. To be honest, launching a product is like giving birth to a baby. All the blood, sweat, and tears that went into making it perfect come down to this one moment. As you launch your product and see it out in the world for the first time, you can't help but feel a sense of pride and accomplishment.

For many companies, the product launch is just the beginning. It's when they start to see if all their efforts have paid off and if the product is actually going to be successful. If the product is successful, it's time to start thinking about scaling and growing it.

Google Wave: Google Wave was a messaging platform that was launched in 2009. It was intended to be an all-in-one messaging platform that would replace email, instant messaging, and online chat. However, the product was unsuccessful and was discontinued in 2010. what went wrong?

There were a number of factors that contributed to the failure of Google Wave. First, the product was too complex, and it was difficult for users to understand how to use it. Second, the product lacked a clear purpose. It was not clear why users would want to use Google Wave instead of other messaging platforms. Third, the product was not well integrated with other Google products. Fourth, the product was not well marketed and did not have a clear target audience.

Like the development stages and their predecessors, the launch stage also has its own set of activities. In fact, the launch stage has the most important set of activities.

What are the activities that are carried out in the launch stage?

1. Verify that the product meets all quality standards and is ready for launch.

2. Create a marketing plan for the product launch.

3. Execute the marketing plan and promote the product to its target audience.

4. Monitor the product's performance after launch and make adjustments as necessary.

As part of the launching efforts, tracking the product's performance and KPIs (key performance indicators) is important. This will help you to determine whether or not the product is successful and identify areas that need improvement.

Some of the KPIs that you may want to track include:

1. Number of users

2. Number of active users

3. Retention rate

4. Engagement rate

5. Conversion rate

6. Revenue

7. Customer satisfaction score

If the product is not performing well, then it may be necessary to go back to the drawing board and make changes. This could involve making changes to the product itself or changing the marketing strategy.

Re-launch the product.

When re-launching a product, it is important to learn from the mistakes made the first time. This will help ensure that the product is more successful the second time.

Some things that you can do to improve your chances of success include:

1. Simplify the product and make it easier to use.

2. Create a clear and concise marketing message that explains the product's purpose and benefits.

3. Identify the product's target audience and tailor the marketing strategy to them.

4. Make sure that the product is well integrated with other products and services.

5. Launch a comprehensive marketing campaign that reaches the target audience.

6. Monitor the product's performance after launch and make adjustments as necessary.

These are just a few things that you can do to improve your chances of success when re-launching a product. If you take the time to learn from your mistakes and make changes accordingly, then you should be able to re-launch the product successfully.

Let's now dive deeper into what documents must be prepared before launching the product. For example, the team must have a product launch plan ready when they launch the product.

Product Launch Plan

Do you know why products fail? Because teams don't launch them with a plan. If you want your product to be successful, you need to have a plan before launching it.

While every company may customize what they include in their launch plan, every launch plan consists more or less of the following elements:

1. A timeline of all the tasks that need to be completed before launch day.

2. A list of all the people who are responsible for each task.

3. A description of the product and its features.

4. The marketing strategy that will be used to promote the product.

5. A budget for the launch.

6. KPIs (key performance indicators) that will be used to track the product's success.

Creating a *launch plan* may seem like a lot of work, but it's essential if you want your product to be successful. Without a plan, it's easy to get overwhelmed and miss important details.

According to a study by the University of Pennsylvania, 43% of product launches fail because the team didn't have a plan. So, if you want your product to be successful, make sure you have a launch plan in place.

Entrepreneur and author Guy Kawasaki has written a helpful guide on how to create a product launch plan. I recommend checking it out if you're not sure where to start.

Now that we've covered the ***basics of product launches,*** let's take a look at some of the common mistakes that teams make.

Mistakes that teams make when launching a product:

1. **Not doing enough market research**: Before you launch a product, it's important to do your homework and understand

the needs of your target market. Without this information, it will be difficult to create a successful product.

2. **Failing to create a comprehensive launch plan**: As we saw earlier, a launch plan is essential for the success of any product. Without one, it's easy to miss important details and fail to meet your goals.

3. **Not having a clear marketing message**: It's important to have a concise and compelling message explaining your product's benefits. If your target audience doesn't understand what your product does, they're unlikely to buy it.

4. **Not promoting the product effectively:** Once you've launched a product, you need to let people know about it. This can be done through various marketing channels such as online advertising, PR, and content marketing.

5. **Not monitoring the product's performance**: It's important to track the performance of your product after launch and make adjustments as necessary. If you see that the product is not meeting your goals, don't be afraid to make changes.

6. **Not being flexible**: The world of business is constantly changing, and you need to be able to adapt to these changes. If you're not flexible, you'll likely fail.

7. **Not learning from your mistakes**: It's important to learn from your mistakes and use this knowledge to improve your

chances of success in the future. If you don't learn from your failures, you're doomed to repeat them.

8. **Giving up too soon**: Launching a product is a long and difficult process, and it's easy to get discouraged. However, it's important to persevere and see the launch through to the end.

9. **Not getting help**: Launching a product is a complex undertaking, and seeking out experts who can help you with specific tasks is often necessary. Don't be afraid to ask for help when you need it.

10. **Trying to do everything yourself**: It's important to delegate tasks and allow others to help you with the launch. Trying to do everything yourself will only lead to burnout and frustration.

If you avoid these common mistakes, you'll be well on your way to launching a successful product.

The elements of a product launch plan vary depending on the product and the team. For now, we have created a launch plan template for you guys:

1. *The product*: A description of the product, including its features and benefits.

2. *The target market*: Who the product is for and why they need it.

3. **The competition**: A description of the competition and how the product is different.

4. **The positioning**: How the product will be positioned in the market.

5. **The messaging**: Key messages that will be used to promote the product.

6. **The launch timeline**: A detailed plan for the launch, including key milestones and deadlines.

7. **The budget**: An estimate of the costs associated with the launch.

8. **The risks**: A list of potential risks that could impact the success of the launch.

9. **The team**: A description of the roles and responsibilities of the launch team.

10. **The measures of success**: Key metrics that will be used to measure the success of the launch.

Following these steps, you can create a comprehensive launch plan to help you achieve your goals.

Some product managers think it wise to make a more elaborate launch plan. Here's a list of that as well.

1. Introduction

2. Goals and objectives

3. the Target market

4. Positioning

5. Pricing

6. Promotions

7. Distribution

8. Timing

9. Budget

10. Metrics

11. Risks and contingencies

12. Conclusion

How do you monitor your product after it is launched?

1. Keep track of your product's KPIs (key performance indicators).

2. Monitor your product's performance on a regular basis.

3. Make adjustments to your product as necessary.

4. Pay attention to customer feedback and make changes based on what you hear.

5. Stay up-to-date on industry trends and make changes to your product accordingly.

6. Be prepared to make changes quickly if your product is not performing well.

7. Always be ready to listen to feedback and make changes based on what you hear.

8. Be willing to experiment with your product and try new things.

After you have launched your product, it is important to keep track of its performance. This can be done by monitoring KPIs (key performance indicators). You should also monitor your product's performance on a regular basis and make adjustments as necessary. Additionally, pay attention to customer feedback and make changes based on what you hear. Finally, stay up-to-date on industry trends and make changes to your product accordingly. By doing these things, you will be able to keep your product up-to-date and improve its performance over time.

Post-Launch:

Now that the product has been launched, it is important to track KPIs to ensure that the product is meeting the target market's needs. Furthermore, surveying the target market at this stage can help you gather feedback about the product and make necessary improvements.

Other activities included in this stage are:

1. Evaluating customer feedback

2. Monitoring sales figures

3. Analyzing customer behavior

4. Conducting marketing campaigns

5. Updating the product based on customer feedback

Many companies, big and small, use customer surveys as one way to collect feedback about their products or services. Surveys are an important part of the product development cycle because they provide insights that can help improve the product before it is launched, and they can also help identify areas for improvement after the product has been launched.

Furthermore, surveys can help companies segment their target market more effectively, identify customer needs and wants, understand buying habits, and track attitudes and perceptions over time. Post-launch activities such as these are important to ensure that the product is meeting customer needs and wants and achieving the desired results.

In post-launch analysis, besides surveys, companies use other techniques like focus groups, interviews, and usability testing to understand customer feedback about their products. All of these methods are important in order to make sure that the product is continuously improving and meeting customer needs. You can think of post-launch analysis as another phase of product development. Some call it "iteration" because it's a never-ending cycle of improvement. Product development never really ends; it just goes on and on and on...

Pro-tip when doing surveys:

1. Keep the questions short and to the point

2. Avoid loaded or leading questions

3. Be clear about the purpose of the survey

4. Give respondents an incentive to participate

5. Keep the survey as anonymous as possible

6. Use a mix of multiple choice and open-ended questions

7. Test the survey before sending it out

8. Use survey software to make analyzing the results easier

Chapter 7

Most Common UX
Design Methods and Techniques

In this chapter, we will discuss:

- Some Common UX design theories and techniques

*Most business models have focused
on self-interest instead of user experience. – Tim Cook*

This section will discuss some amazing, simple, and effective methods and UX programming theories for creating amazing products. Since our focus is on creating products, we also need to look at things from a business perspective. UX designers in the industry have used these methods for a long time.

Irrespective of what product you are creating, these points will help make the product creation cycle and process much easier.

Value Statement

A value statement is a simple, one- or two-sentence description of the core benefit that your product or company offers. It should be clear, concise, and easy to remember.

For example:

- "Our product helps companies save time and money."

- "Our product helps people stay healthy and fit."

- "Our product makes it easy to find the perfect gift."

A value proposition is similar to a value statement, but it is more specific. It answers the question, "Why should someone buy our product or use our service?"

For example:

"Our product is the fastest and most efficient way to find the perfect gift."

By defining your *value statement* or *proposition* early on, you can make sure that your product is designed to deliver the promised value. How does the value proposition of a product help a UX designer?

When you have a clear value proposition, it becomes easier to determine what features and functions your product needs to have. It also helps you focus on the user's needs and how your product can meet them.

Let us help you understand the value proposition in UX by giving you an example:

Suppose you are designing a new email service. What features would you include?

Some possible features could be:

- The ability to send and receive email
- The ability to create folders to organize emails
- The ability to search for specific emails
- The ability to add attachments to emails
- The ability to set up email filters

Now, let's say that your value proposition is "Our email service helps people stay organized and save time." With this in mind, you would likely prioritize features that help the user stay organized, such as the ability to create folders and set up email filters.

Your Product Strategy

After you have determined your product's value proposition, you need to develop a product strategy. This will help you determine what features and functions your product needs to have and how you will bring your product to market.

Your product strategy should answer the following questions:

- What problem does your product solve?
- Who is your target market?
- What are your product's key features and benefits?
- How will you position your product in the market?
- What is your go-to-market strategy?

By answering these questions, you will clearly understand what your product needs to do and how you will sell it. This will help you make better product decisions and avoid making costly mistakes.

Creating Buyer Personas

A buyer persona is a fictional character that represents your ideal customer. When creating a buyer persona, you should consider your customer's demographics, needs, goals, and pain points.

Creating a buyer persona can help you understand your target market and make better product decisions. It can also help you create more effective marketing campaigns.

To create a buyer persona, you can use the following template:

Name: _____

Demographics: _____

Needs: _____

Goals: _____

Pain points: _____

Once you have created your buyer persona, you can use it to help you make decisions about your product. For example, if your persona's goal is to save time, you would likely prioritize features that help the user save time.

Heuristics Evaluation

A heuristic evaluation is a method of usability testing where a team of experts evaluates a product against a set of criteria or heuristics. This type of evaluation is typically used to identify usability issues with a product.

To conduct a heuristic evaluation, you will need to select a team of experts and provide them with a list of criteria to evaluate your product against. The experts will then use the product and provide feedback on any areas that need improvement.

Heuristic evaluations are a great way to identify usability issues with a product. However, they can be expensive and time-consuming.

Example of heuristic evaluation:

- Select a team of experts.
- Provide the experts with a list of criteria to evaluate your product against, such as the following:
- Is the product easy to use?
- Is the product well designed?
- Does the product meet the user's needs?

- Is the product usable on all devices?

- Is the product accessible to all users?

- The experts will use the product and provide feedback on any areas that need improvement.

- Based on the experts' feedback, you can make changes to improve the usability of your product.

Meeting with Stakeholders

Stakeholders are people who have a vested interest in your product. They can be customers, investors, employees, or anyone else who stands to gain or lose from your product's success or failure.

Meeting with stakeholders is a great way to get feedback on your product and ensure that you meet their needs. It can also help you build relationships and earn trust.

To have a successful meeting with stakeholders, you should:

- Prepare an agenda in advance.

- Invite the right people to the meeting.

- Be clear about what you want to achieve.

- Be open to feedback and willing to make changes.

- Follow up after the meeting.

By following these tips, you can ensure that your meeting is productive and that you are making the right decisions for your product.

Conducting User Research

User research is the process of gathering data about your users. This data can be used to understand user needs, identify usability issues, and make better product decisions.

User research can be conducted in a variety of ways, such as interviews, surveys, focus groups, and usability testing. The best method of user research will depend on your specific needs.

User research is an essential part of the product development process. By conducting user research, you can ensure that you are making decisions that are based on real data, not assumptions.

Techniques used to inspire ideas in a design process:

Brainstorming:

Brainstorming is a technique used to generate new ideas. It involves a group of people working together to come up with ideas. Brainstorming can be used in various situations, such as when you're trying to come up with a new product or trying to solve a problem.

To brainstorm effectively, you should:

- Set a goal for the brainstorming session.
- Invite a diverse group of people to participate.
- Encourage wild and crazy ideas.
- Don't criticize any ideas.
- Combine and build on ideas.

Cultural Probes

Cultural probes are a type of user research. They involve giving users a kit of materials to use and then asking them to document their experiences. The kit can include a diary, a camera, and other materials.

Cultural probes are a great way to gather data about how users interact with your product in their everyday lives. They can also be used to generate new ideas for product development.

To use cultural probes effectively, you should:

- Give users clear instructions.

- Make sure the materials you provide are user-friendly.

- Be prepared to receive a lot of data.

- Analyze the data carefully.

Ethnographic Research

Ethnographic research is similar to cultural probes. Ethnographic research is a type of user research that involves observing users in their natural environment. This type of research is often used to understand how users interact with a product in their everyday lives.

Like cultural probes, ethnographic research can be used to generate new ideas for product development. It can also be used to understand user needs and identify usability issues.

To do ethnographic research effectively, you should:

- Choose the right location.

- Observe users without being intrusive.

- Take detailed notes.

- Analyze the data carefully.

Focus Groups

Focus groups are a type of user research. They involve bringing together a group of people to discuss a topic. Focus groups can be used to gather data about user needs, identify usability issues, and generate new ideas for product development.

To run a successful focus group, you should:

- Prepare an agenda in advance.

- Invite a diverse group of people to participate.

- Moderate the discussion.

- Encourage participants to speak freely.

- Take detailed notes.

Your Roadmap of the Product

A product roadmap is a document that outlines the plans for a product. It includes a timeline of when features will be developed and released and an overview of the product's features.

A product roadmap is an essential tool for product development. It helps you to plan and track the development of your product. It also helps you to communicate your plans to stakeholders.

To create a product roadmap, you should:

- Define the goals for your product.
- Identify the features that will help you to achieve those goals.
- Prioritize the features.
- Create a timeline for development.
- Communicate the roadmap to stakeholders.

Once you have created your product roadmap, you should review it regularly and update it as needed.

Prototyping

Prototyping is a method of creating a model of a product. Prototypes can be used to test how users interact with a product and to identify usability issues.

Prototyping is an essential part of product development. It helps you to test your ideas and to make sure that your product is user-friendly.

To create a prototype, you should:

- Identify the purpose of the prototype.

- Choose the right type of prototype for your needs.

- Make sure the prototype is user-friendly.

- Test the prototype with users.

When you are prototyping, it is important to keep in mind that the prototype is not the final product. The goal is to use the prototype to gather data and improve the design of the product.

Summative Evaluation

Summative evaluation is a type of user research. It involves assessing a product after it has been released. Summative evaluation can be used to identify usability issues, gather user feedback, and assess the overall success of a product.

To do a summative evaluation, you should:

- Choose the right type of evaluation for your needs.

- Set up the evaluation process.

- Collect data from users.

- Analyze the data.

When you are doing a summative evaluation, it is important to remember that the goal is to improve the product. The data you collect should be used to make product changes and improve the user experience.

Task Breakdown

A task breakdown is a list of the steps that are needed to complete a task. It is used to help you understand how a task is done and to identify potential problems.

A task breakdown is an essential tool for product development. It helps you to understand the user experience and to identify potential usability issues.

To create a task breakdown, you should:

- Identify the task that you want to break down.

- List the steps involved in the task.

- Identify any potential problems.

- Test the task breakdown with users.

When creating a task breakdown, it is important to remember that the goal is to improve the user experience. The task breakdown should be used to identify potential problems and make product changes.

User Story

A user story is a description of how a user will use a product. User stories are used to help you understand the user experience and to identify potential usability issues.

User stories are an essential tool for product development. They help you to understand the user *experience and to identify potential problems.*

147

To create a user story, you should:

- Identify the user.

- Describe how the user will use the product.

- Identify any potential problems.

When creating a user story, it is important to remember that the goal is to understand the user experience. The user story should be used to identify potential problems and make product changes.

SWOT Analysis

A ***SWOT analysis*** is a tool used to assess a product's strengths, weaknesses, opportunities, and threats. It can be used in a variety of situations, such as when you're trying to decide whether to launch a new product or improve an existing product.

To do a SWOT analysis, you should:

- Identify the strengths of your product.

- Identify the weaknesses of your product.

- Identify the opportunities for your product.

- Identify the threats to your product.

After you've identified the SWOT elements, you should analyze them and use them to make decisions about your product.

A case study of SWOT analysis:

Company X is considering launching a new product. They use a SWOT analysis to assess the potential of the product.

Strengths:

- The product is unique, and there is no competition.

- The product has a high potential to be successful.

Weaknesses:

- The product is still in development and has not been tested.

- The company does not have a lot of experience with this type of product.

Opportunities:

- The product has the potential to be very successful.

- There is a lot of potential for growth in this market.

Threats:

- The product may fail if it is not executed well.

- The competition may copy the product and launch a competing product.

After doing the SWOT analysis, the company decides to launch the product. They believe that the product's strengths outweigh the weaknesses and that there is a lot of potential for the product to succeed.

Competitive Analysis

A competitive analysis is a tool used to assess the strengths and weaknesses of your competitors. It can be used in a variety of situations, such as when you're trying to decide whether to enter a new market or when you're trying to improve your position in an existing market.

To do a competitive analysis, you should:

- Identify your competitors.

- Assess the strengths and weaknesses of your competitors.

- Analyze the data and use it to make decisions about your product.

User Testing

User testing is a type of user research that involves testing a product with real users. User testing can assess a product's usability, identify user needs, and generate new ideas for product development.

To do user testing effectively, you should:

- Identify the goals of your user test.

- Select the right participants.

- Prepare a script.

- Conduct the test.

- Analyze the results.

After you've conducted the user test, you should analyze the results and use them to make decisions about your product.

A case study of user testing:

Company X is developing a new website. They want to make sure that the website is easy to use, so they decide to do some user testing.

They identify the goals of their user test, which are to assess the usability of the website and to identify user needs. They select the right participants for their test, which are people who are representative of their target audience. They prepare a script for the test, which includes tasks that the participants will need to complete. They conduct the test and analyze the results.

Based on the results of their user test, they made some changes to the website. They add a search function and make navigation more intuitive. They also add some features that their users have requested.

In conclusion, UX programming theories can be used to design great products. These theories can be used to assess the strengths and weaknesses of your product, identify user needs, and generate new ideas for product development. Using these theories, you can create products that are easy to use and meet your users' needs.

The most common case study that people use to understand the product development process is the creation of the iPhone. Apple went through all stages of product development when creating the iPhone. They started with user research to understand the needs of their users. They then created prototypes and tested them with users. They then created the final product and released it to the market. After the product was released, they gathered feedback from users and used it to improve the product. The iPhone is a great example of how user experience can enhance a brand. The iPhone was easy to use and had a sleek design that was different from other phones on the market. The user experience was so good that people were willing to pay more for the iPhone than for other phones. The iPhone helped Apple establish itself as a premium brand.

Let us take the example of a product that goes through stages such as Research and Planning, Design, Development, Testing and launch and Post-launch analysis/Iterate:

To help you show how a product goes through all these stages of development, let us take the example of a laptop.

The first stage of developing a laptop would be research and planning, wherein the company decides what they want to make, how it should look, and the specs it should have. They would also look into their competition and try to figure out how to make their product better. The team would brainstorm and use tactics such as user interviews and surveys to understand the needs of their target market.

After they have done their research and planning, they would then move on to the design stage. In this stage, they would create prototypes. the design team can use various software such as Sketch or Photoshop to create their designs. Users would test these prototypes to see if they could use the product easily and if they liked the design. After getting feedback, the team would make necessary changes and improve the design.

Side by side, the team would be creating documentation such as user flows and sitemaps. This would help them keep track of the different screens in the product and how users would move from one screen to another.

Once the design is finalized, the team will start development. In this stage, they would take all the designs and turn them into code. They would also create the backend infrastructure and integrate different third-party services. The team will test the product to see if everything works as intended. If there are any bugs, they will be fixed at this stage.

After the product is developed and tested, it is ready to launch. The team would create marketing materials and set up a website at this stage. They would also decide on a pricing strategy and launch the product on different platforms.

After the product is launched, the team would start gathering user feedback. They would use this feedback to improve the product and make necessary changes. This process is known as iteration.

As you can see, there are many stages that a product goes through before it is released to the market. Creating a great product requires a team of experienced professionals who are familiar with all

Practical Task

Creating User Flow of a Product:

1. start by creating a list of all the steps a user would take when using your product

2. then, create a visual representation of the user flow by creating a diagram or flowchart

3. finally, add annotations to your diagram to explain each step in the user flow

Here's an example of a simple user flow for a hypothetical e-commerce product:

1. The user arrives on the home page and is presented with a list of products

2. The user clicks on a product that interests them

3. The user is taken to the product page, where they can learn more about the product

4. The user adds the product to their cart and proceeds to checkout

5. The user enters their shipping information and payment details

6. The user confirms their order and is taken to the confirmation page

7. The user is redirected to the home page, and the cycle begins anew

Create a Product Roadmap of a Hypothetical Mobile App Called "Yum":

1. The app will allow users to find and share recipes

2. Users will be able to search for recipes by ingredient, dish type, or dietary restriction

3. Users will be able to save their favorite recipes for easy access later

4. Users will be able to share recipes with their friends via social media or email

Here's how I started it. You can fill in the rest and continue the journey.

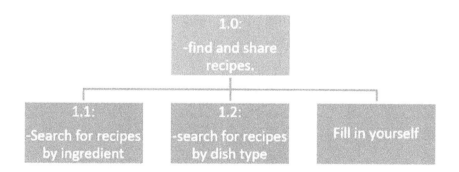

Conclusion

We hope this book has given you a better understanding of using UX programming theories to design great products. As we said before, these theories can help you understand how you can design BETTER and more effective products.

We covered the 5-step product process chapter by chapter. We gave case studies and pro tips, so you may relate to how these theories work in real life. Since this book focuses on making great products using UX theories, we consider it imperative to mention how user experience can enhance your brand and what it means for your business. This idea is covered in chapter 1.

Chapter 2 briefly covers the product phases, and the documentation created dusting each product stage. In the last chapter, we look back at some UX theories and methods or techniques that designers swear by and are imperative to a product's success.

Using these theories, you can create products that are easy to use and meet your users' needs. Furthermore, a good product design will reflect the company's brand and enhance its reputation. Creating a great product requires more than just using the latest UX theories and methods. It also requires careful planning, execution, and ongoing user feedback. But with the right approach, you can create products that are not only easy to use but also meet the needs of your users.

So what are you waiting for? Get out there and start using these theories to create amazing products!

Thank you for buying and reading/listening to our book. If you found this book useful/helpful please take a few minutes and leave a review on Amazon.com or Audible.com (if you bought the audio version).

References

Babich, N. (2020, October 16). *Most Common UX Design Methods and Techniques - UX Planet*. Medium. https://uxplanet.org/most-common-ux-design-methods-and-techniques-c9a9fdc25a1e

Buxton, B. (2007). *Sketching User Experiences: Getting the Design Right and the Right Design (Interactive Technologies)* (1st ed.). Morgan Kaufmann.

USER EXPERIENCE (UX) DESIGN CONCEPTS FOR MOBILE APP DEVELOPMENT COURSES. (2020). *Issues In Information Systems*. https://doi.org/10.48009/4_iis_2020_202-211

Allabarton, R. (2022, January 7). *What Is the UX Design Process? A Complete, Actionable Guide*. CareerFoundry. https://careerfoundry.com/en/blog/ux-design/the-ux-design-process-an-actionable-guide-to-your-first-job-in-ux/

5 Principles of Visual Design in UX. (2020, March 1). Nielsen Norman Group. https://www.nngroup.com/articles/principles-visual-design/

Spencer, J. (2020, October 23). *7 Ways UX Design Theory Transformed My Approach to Course Design*. John Spencer. https://spencerauthor.com/ux/

Press, Scripto Love. (2021). *UX/UI Designer Notebook (White): UX/UI Design for Mobile, Tablet, and Desktop - Sketchpad - User Interface - Experience App Development - Sketchbook - . . . App MockUps - 8.5 x 11 Inches With 120 Pages*. Independently published.

Gothelf, J., & Seiden, J. (2021). *Lean UX: Designing Great Products with Agile Teams* (3rd ed.). O'Reilly Media.

Bassino, N. (2021). *Product Direction: How to build successful products at scale with Strategy, Roadmaps, Objectives and Key Results (OKRs)*. Nacho Bassino.

Bassino, N. (2021). *Product Direction: How to build successful products at scale with Strategy, Roadmaps, Objectives and Key Results (OKRs)*. Nacho Bassino.

Lombardo, T. C., McCarthy, B., Ryan, E., & Connors, M. (2017). *Product Roadmaps Relaunched: How to Set Direction while Embracing Uncertainty* (1st ed.). O'Reilly Media.

Selikoff, S. (2020). *The COMPLETE BOOK of Product Design, Development, Manufacturing, and Sales*. Independently published.

Jones, C. (2021). *UX/UI Design 2022: A Comprehensive UI & UX Guide to Master Web Design and Mobile App Sketches for Beginners and Pros*. Independently published.

www.ingramcontent.com/pod-product-compliance
Lightning Source LLC
Chambersburg PA
CBHW071159050326
40689CB00011B/2179

9 781088 225707